Questions and Answers
O Level Physics

Questions & Answers

O LEVEL PHYSICS

R.H. Evans
B.A., B.Sc.

Checkmate/Arnold

© R.H. EVANS 1986

First published in Great Britain 1986 by
Checkmate Publications,
4 Ainsdale Close, Bromborough, Wirral L63 0EU.

This edition published in association with
Edward Arnold (Publishers) Ltd.,
41 Bedford Square, London WC1B 3DQ

Edward Arnold (Australia) Pty Ltd.,
80 Waverley Road, Caulfield East,
Victoria 3145, Australia

Edward Arnold, 3 East Read Street,
Baltimore, Maryland 21202, USA.

All rights reserved. No part of this publication may be reproduced or transmitted in any form or by any means, without the written permission of the Publisher.

ISBN 0 946973 25 3

Printed and bound by Richard Clay (The Chaucer Press),
Bungay, Suffolk

Introduction

This book is intended to help students about to take examinations at the G.C.E. 'O' level in Physics. The questions are drawn from both the University of London and the Associated Examining Board's papers. These Boards have given their kind permission for reproduction of their questions in this text, however we must point out that this in no way implies that the solutions given are the responsibility of either Board. The solutions are the sole responsibility of the author.

The format of the book is specifically structured so that students may read and attempt questions before referring to the suggested answer. In this respect it is a useful self testing program.

The author of this text is a long standing member of the teaching profession specialising in Mathematics and Physics. Prior to entering the profession many years ago he was an engineer and thus his experience in the applied field is invaluable.

Table of Contents

Number	Topics	Page No.
1	Conduction of heat, linear expansion, energy	1
2	Specific heat capacity, specific latent heat. Conversion of electrical energy to heat energy	2
3	Evaporation, boiling, refrigeration	3
4	Saturated vapour pressure, boiling	5
5	Thermometry, refrigeration	6
6	Specific heat capacity of a liquid	8
7	Velocity, acceleration, kinetic energy	9
8	Gravitation, potential energy, kinetic energy	10
9	Heat, electrical, potential and kinetic energies. Power	11
10	Speed-time graph	14
11	Density and relative density	15
12	Archimedes Principle. Pressure	16
13	Molecular theory. Fusion	16
14	Simple pendulum. Wave motion	17
15	Sound waves. Vibrating strings	18
16	Sound waves. Tuning fork	20
17	Ray diagram:- convex lens. Simple camera	21
18	Refraction of light. Speed and wavelength	23
19	Reflection and total internal reflection of light	24
20	Refraction of light	25
21	Experimental work involving a lens	27
22	Series and parallel connection of resistors. Simple electrical circuits	28

Number	Topics	Page No.
23	Voltmeter. Internal resistance of cells	29
24	Electro-magnetism. Simple electric motor/generator	31
25	Simple electrical circuits, electrical energy	33
26	Experimental determination of resistance. Thermistor	34
27	Electro-magnetic induction	35
28	Detection of radioactive emission	37
29	Cathode ray tube. Radioactivity	39
30	Positive and negative charges	40

Question 1 (answer - page 42)

PART I

(a) When a short piece of hollow metal pipe is heated in a bunsen burner flame, what changes, if any, occur to its

   (i) length
   (ii) external diameter
   (iii) internal diameter? (3 marks)

(b) Using such a pipe, how would you demonstrate thermal conduction in the laboratory? (5 marks)

PART II

(c) The pipe described in Part I is made from lead and, at 290 K, has an internal diameter of 6 cm and an external diameter of 10 cm. The cross-sectional area of lead may be assumed to be 50 cm^2. If the mass of the pipe is 22 kg find the length of the pipe at 290 K. (3 marks)

(d) The pipe is heated in a furnace from 290 K to 490 K. Determine

   (i) the increase in length of the pipe
   (ii) the increase in thickness of the wall of pipe
   (iii) the energy supplied to the pipe to raise its temperature from 290 K to 490 K.

   (7 marks)

(e) If it takes as much energy to melt a given mass of lead as it does to raise the temperature of the same mass of lead through 200 K, calculate the specific latent heat of fusion of lead. (3 marks)

Data for this Question:

density of lead at 290 K = 11 000 kg/m^3
specific heat capacity of lead = 125 J/(kg K)
linear expansivity of lead = 3 x 10^{-5}/K

(A.E.B. Nov 1984)

Question 2 (answer - page 43)

PART I

(a) Define specific heat capacity and specific latent heat. (4 marks)

(b) 1 kg of water is heated at a constant rate and initially the temperature rises at a rate of 5 K per minute. The rate of rise of temperature decreases until the temperature is 373 K. The temperature then remains constant. An equal mass of glycerol is heated by the same source under the same conditions and the temperature rises initially at the rate of 10 K per minute. The temperature of the glycerol ceases to rise at 563 K.

Explain these observations. (4 marks)

PART II

(c) When a current of 4 A is passed through a 6 Ω coil immersed in 0.25 kg of water the temperature of the water rises from $20\,^{\circ}\text{C}$ to $25\,^{\circ}\text{C}$ in a time of 1 minute.

Determine

  (i) the potential difference across the coil

  (ii) the power output of the heating coil

  (iii) the energy supplied to the heating coil in 1 minute

  (iv) the average rate of loss of energy to the container and surroundings while the temperature of the water is rising from 20^{0}C to 25^{0}C.
  (10 marks)

(d) In a second experiment the heating coil is used to raise the temperature of the water to its boiling point. During the experiment it is noted that the current through the coil decreases even though the potential difference is kept constant. How would you explain this decrease? (3 marks)

CONT. OVER......

3.

Data for this Question:

specific heat capacity of water = 4200 J/(kg K)

(A.E.B. June 1984)

Question 3 (answer - page 44)

PART I

(a) Name **two** factors which affect the boiling point of water, and in each case state how the boiling point is affected. (2 marks)

(b) Distinguish between evaporation and boiling. (4 marks)

(c)

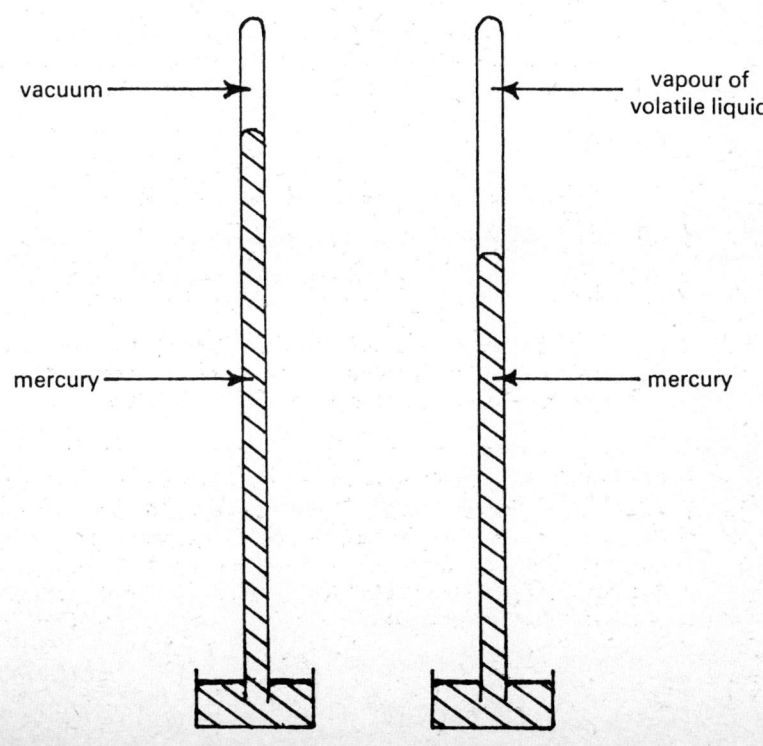

The diagrams show a glass tube before and after a small amount of a volatile liquid has been introduced into the space above the mercury. Explain why the mercury in the right hand tube is lower than that in the left hand tube. (2 marks)

PART II

(d)

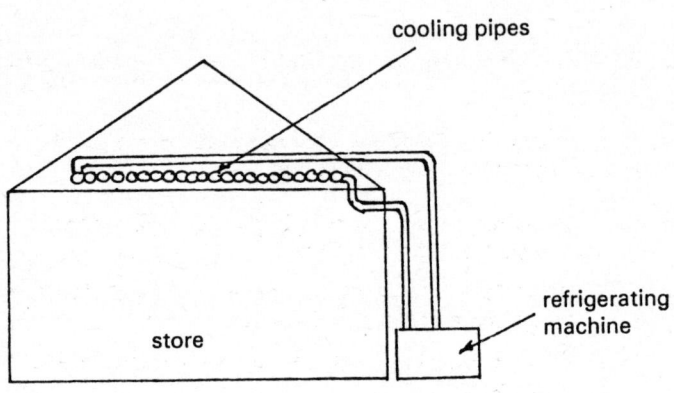

The diagram shows the essential parts of a cold store.

    (i)   Why are the cooling tubes in the roof?

    (ii)   Explain carefully why the refrigerating machine is outside the store.

    (iii) Why, even if the refrigerating machine runs continuously, do the contents not get colder and colder? (6 marks)

(e) A copper block of mass 0.68 kg is suspended in a freezing mixture at $-50°C$ for some time and then transferred to a large volume of water at $0°C$. A layer of ice is formed on the block.

    (i)   Explain why the ice is formed.

    (ii)   What will be the temperature of the copper block after this change is complete?

    (iii) Calculate the mass of ice formed. (7 marks)

CONT. OVER......

5.

Data for this Question:

Specific heat capacity of copper = 400 J/(kg K)
Specific latent heat of ice = 340 000 J/kg

(A.E.B. Nov 1982)

Question 4   (answer - page 45)

PART I

(a) Explain what is meant by the saturated vapour pressure of a liquid.   (2 marks)

(b) A strong-walled flask which contains some warm water is connected to a vacuum pump.

   (i) Why does the water begin to boil when air is pumped out of the flask?

   (ii) Why does the water continue to boil for a short time after the pump is switched off?

   (iii) Why does the temperature of the water fall as it boils?   (6 marks)

PART II

(c) Variation of saturated vapour pressure of water with temperature

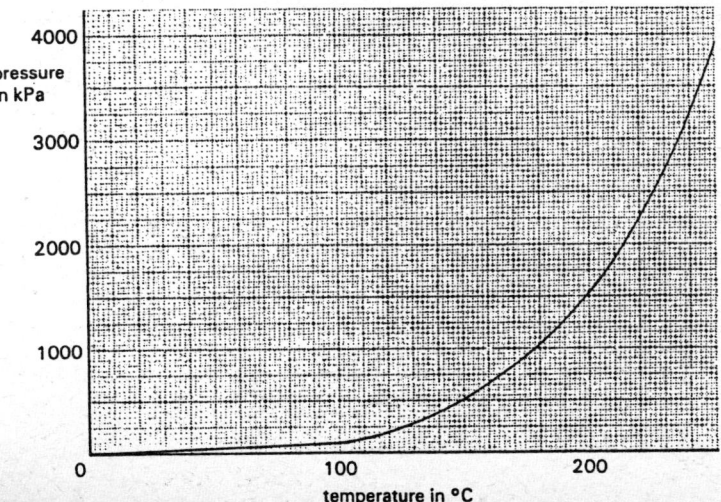

The graph shows the variation of the saturated vapour pressure of water with temperature.

(i) What temperature corresponds to a saturated vapour pressure of 100 kPa (i.e. atmospheric pressure)?

(ii) To what value must the pressure on water be increased to raise its boiling point to 200^0 C?

(iii) The steam pressure in an industrial boiler is 1200 kPa. What is the temperature of the steam in the boiler? (3 marks)

(d) A laboratory distillation apparatus produces 0.1 kg of pure water per minute. Cooling water flows through the apparatus at the rate of 1 kg per minute.

(i) Calculate the energy evolved per minute by the steam condensing and cooling to 70^0C.

(ii) If the initial temperature of the cooling water is 5^0 C, what is its temperature on leaving the apparatus? State any assumption you have made.
(10 marks)

Data for this Question:

Specific heat capacity of water = 4200 J/(kg K).
Specific latent heat of vaporisation of water = 2.26 MJ/kg.

(A.E.B. June 1983)

Question 5   (answer - page 47)

PART I

(a) Draw a labelled diagram to show the essential features of a $-10\ ^0$C to 110^0 C mercury-in-glass thermometer. Mark on the diagram the position of the fixed points and their values. (4 marks)

(b) How are the boiling point and freezing point of water affected by

(i) the presence of dissolved solids

(ii) changes in atmospheric pressure? (4 marks)

PART II

(c) The diagram shows a capillary tube containing a small volume of dry air trapped by a pellet of mercury.

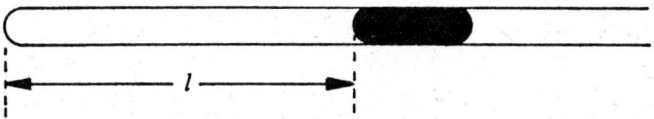

The length, l, of the air column is measured at various temperatures. The results obtained are shown in the table.

Conditions	l in mm
At lower fixed point	128
At upper fixed point	178
At room temperature	136

(i) What is room temperature in degrees Celsius?

(ii) What would be the steady length of the air column when the tube is in a freezer at $-10°C$?

(4 marks)

(d) Explain, with the aid of a diagram of a refrigerator, how the effect of pressure on the boiling point of a liquid is used in a refrigerating system. (6 marks)

(e) Why is the cooling unit of a refrigerator located near the top of the cabinet? (3 marks)

(A.E.B. Nov 1983)

Question 6 (answer - page 49)

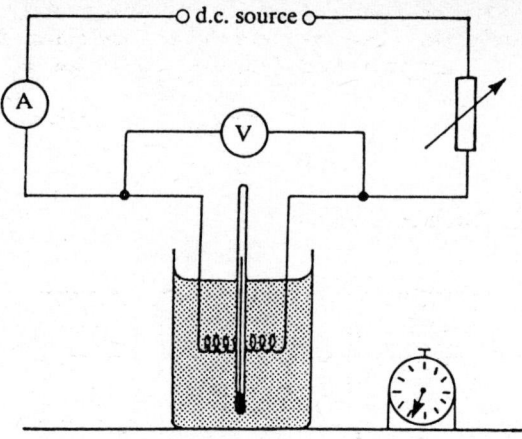

The diagram illustrates an experiment in which the electrical energy used to produce a measured rise in temperature of a liquid can be determined. Explain why

(a) the liquid will tend to be warmer at the top than at the bottom of the container, (3 marks)

(b) the rate of rise of temperature would be increased if the container were

    (i)   lagged with glass fibre,
    (ii) covered with a lid, (4 marks)

(c) the temperature will eventually stop rising even though the current is still passing through the heating coil,
(3 marks)

(d) if the apparatus is used to determine the specific heat capacity of the liquid, the accuracy of the experiment will be increased if the liquid is first cooled to about 5 K below room temperature and the current passed until the temperature is about 5 K above room temperature. (3 marks)

In the above apparatus the heater supplies 36 J/s to a liquid of mass 0.5 kg and specific heat capacity 4200

J/kg K. Calculate the rate of rise of temperature in K/min. (Neglect the effect of the container.) (5 marks)

If allowance were made for the container, what effect, if any, would this have on the value calculated for the rate of rise of temperature? (Explain your answer.)
(2 marks)
(U.O.L. Jan 1981)

Question 7   (answer - page 50)

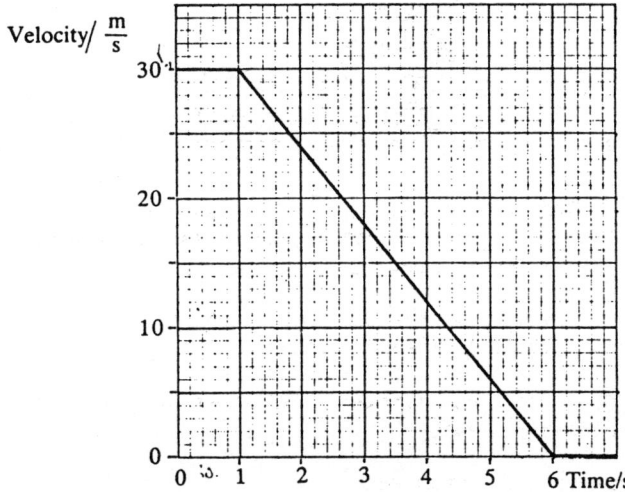

The graph shows how the velocity of a car of mass 1500 kg varies with time after the driver first sees an obstruction on the road 106 m ahead. The driver has a reaction time of 1.0 s between seeing the obstruction and first applying the brakes.

(a) Use the graph to determine whether or not the car collides with the obstruction. (Show clearly how you arrive at your answer.) (4 marks)

(b) Determine the retardation and hence the braking force experienced by the car while it is slowing down.
(4 marks)

(c) Explain why the driver tends to be thrown forward in the car as it slows down. (4 marks)

(d) Calculate the kinetic energy of the car when travelling at a velocity of 30 m/s. (3 marks)

(e) Explain why, if the car had been travelling at 15 m/s, the braking distance would have been greater than 1/4 of the braking distance when the velocity was 30 m/s. (You may assume that the same braking force was applied.) (5 marks)

(U.O.L. Jan 1981)

Question 8 (answer - page 51)

PART I

(a) A piece of paper and an apple are dropped from a high cliff. Explain briefly why

   (i) the piece of paper takes longer to reach the ground than the apple

   (ii) if the apple and paper were dropped through the same height above the surface of the moon, the time taken for the apple and the piece of paper to reach the surface would be the same. (4 marks)

(b) If on the moon the apple reaches a speed of 5 m/s in a time of 3 s from rest, what is the acceleration due to gravity at the surface of the moon? (4 marks)

PART II

(c) An aeroplane flying at a constant height of 320 m drops a smooth dense object of mass 100 kg. Determine, assuming that there are no frictional losses,

   (i) the loss in potential energy of the object on falling to the ground

   (ii) the gain in kinetic energy of the object on falling to the ground

(iii) the vertical component of the speed with which the object hits the ground. (5 marks)

(d) The same aeroplane, a short time later, drops a load fitted with a parachute. During the time intervals stated, the motion of the load is as follows:

    0 s - 2 s   free fall, negligible resistance

    2 s - 6 s   parachute opens, vertical speed decreases uniformly to 2 m/s

    after 6 s   the load falls with a constant vertical speed of 2 m/s.

(i) Sketch a speed-time graph for the first 8 s.

(ii) Determine the distance fallen in the first 6 s.

(iii) Determine the total time taken for the load to reach the ground. (8 marks)

Data for this Question:

acceleration due to gravity, $g = 10$ m/s^2

(A.E.B. June 1984)

Question 9 (answer - page 53)

PART I

(a) Write down a formula for each of the following:

(i) the potential energy of an object of mass m at a height h above sea level,

(ii) the kinetic energy of an object of mass m moving with a speed of v,

(iii) the electrical energy used by an immersion heater of resistance R carrying a current of I for a time t,

(iv) the heat given out when a solid of mass m and specific heat capacity s cools through a temperature drop of θ. (4 marks)

(b)

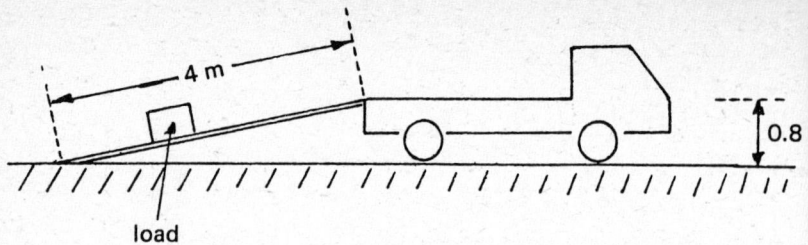

load

The diagram shows a loading ramp being used to raise a load of weight 2000 N on to a lorry.

(i) How much work is done against gravity in raising the load?

(ii) Explain why the force required to move the load up the ramp is less than 2000 N.

(iii) Explain why the work done in moving the load up the ramp is greater than the work done against gravity. (4 marks)

PART II

(c) A catapult is used to fire a stone of mass 50 g vertically to a height of 4.05 m. Calculate

(i) the potential energy gained by the stone,

(ii) the speed of the stone as it leaves the catapult.

(6 marks)

continued......

(d)

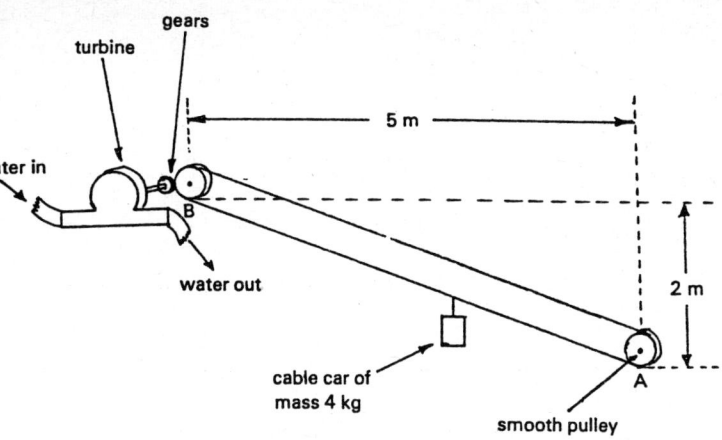

The diagram shows a driving mechanism consisting of a water turbine and a set of gears. It is being used to lift a model cable car from A to B in 20 s.

Calculate

(i) the potential energy gained by the cable car in going from A to B,

(ii) the **output** power of the driving mechanism,

(iii) the **input** power to the turbine, if the driving mechanism is 80% efficient. (7 marks)

Data for this Question:

Acceleration due to gravity = 10 m/s^2

(A.E.B. Nov 1982)

Question 10   (answer - page 54)

The speed of a train which is hauled by a locomotive varies as shown below as it travels between two stations along a straight horizontal track.

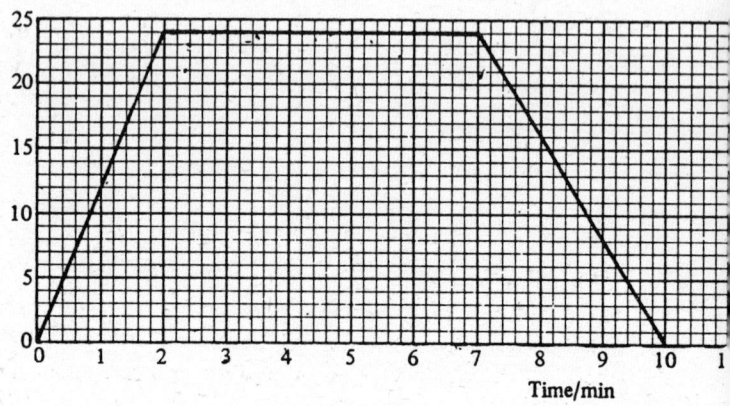

Use the graph to determine

(a) the maximum speed of the train,
(b) the acceleration, in m/s², of the train during the first two minutes of the journey,
(c) the time during which the train is slowing down,
(d) the total distance, in metres, between the two stations along the line,
(e) the average speed, in m/s, of the train.   (14 marks)

Explain why the pull of the locomotive on the train during the first two minutes of the journey must be greater than the pull during the next five minutes of the journey.
(3 marks)

Is there any part of the journey when the pull of the locomotive is equal to the resistive forces acting on the train? Explain your answer.   (3 marks)

(U.O.L. June 1982)

Question 11 (answer - page 55.)

(a) Distinguish between the density and the relative density of a substance. State the units in which density and relative density may be quoted. (4 marks)

(b) In order to determine the relative density of a liquid, a specially made small bottle may be used.

   (i) How is the bottle designed to enable identical volumes of liquid to be taken each time it is used? (3 marks)

   (ii) Explain why it is inadvisable to hold the bottle closely in the hand when making such a determination? (2 marks)

   (iii) List the readings that should be taken and show how they would be used to determine the relative density of the liquid. (4 marks)

(c)

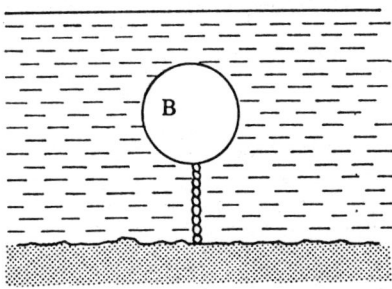

A body B of mass 200 kg is anchored, as shown, to the bottom of a lake by a light chain. The tension in the chain is 500 N. Calculate the mass of water displaced by the body. (4 marks)

If the body is moved into salty water of the same depth, will the tension in the chain increase or decrease? (Explain your answer carefully.) (3 marks)

(U.O.L. Jan 1981)

Question 12   (answer - page 56)

PART I

(a) State Archimedes' Principle. (2 marks)

(b) (i) Draw a labelled diagram of a bulb and stem hydrometer.

   (ii) What physical quantity does a hydrometer measure?
   (4 marks)

(c) Define pressure and write down the formula used to determine the pressure, in Pa, due to a column of liquid. State the meaning of any symbols used.
(2 marks)

PART II

(d) A cube of wood of volume 0.2 m^3 and density 600 kg/m^3 is placed in a liquid of density 800 kg/m^3.

   (i) What fraction of the volume of the wood would be immersed in the liquid?

   (ii) What force must be applied to the cube so that the top surface of the cube is on the same level as the liquid surface? (6 marks)

(e) Describe, with the aid of a labelled diagram, how you would measure the excess pressure, in Pa, of the laboratory gas supply, using a manometer.
(7 marks)

Data for this Question:

Acceleration due to gravity = 10 m/s^2

(A.E.B. Nov 1982)

Question 13   (answer - page 58)

(a) Describe differences in the behaviour of the molecules of a substance in its solid, liquid and gaseous states.
(3 marks)
Using these differences as a basis, explain why

   (i) energy must be removed from a liquid at its

freezing point for it to solidify, (2 marks)

(ii) evaporation of a liquid may produce a cooling of the liquid. (5 marks)

Explain why the pressure of an enclosed gas at a constant temperature increases when the volume is reduced.

(b) It is required to convert 0.5 kg of water at $20\,^0C$ into ice using a refrigerator which can extract heat at an average rate of 20 J/s. Determine whether this is possible within a period of 2 h.

(Specific heat capacity of water = 4200 J/kg K.
Specific latent heat of fusion of ice = 336 000 J/kg.)
(8 marks)
(U.O.L. June 1982)

Question 14 (answer - page 60)

PART I

(a) A simple pendulum has a period of oscillation of 1.5 s and an amplitude of oscillation of 6 cm.

   (i) Sketch a simple diagram of the pendulum showing what is meant by the term amplitude of oscillation.

   (ii) What is meant by the period of oscillation?

   (iii) What is the frequency of oscillation of the pendulum?

   (iv) Sketch, in your answer book, a displacement-time graph for the motion of the pendulum over a time of 3 s. Label the graph as fully as possible and mark on it the period and the amplitude of oscillation. (8 marks)

PART II

(b) A vibrating tuning fork is placed on a sonometer board and produces a standing wave like the one shown in the diagram below.

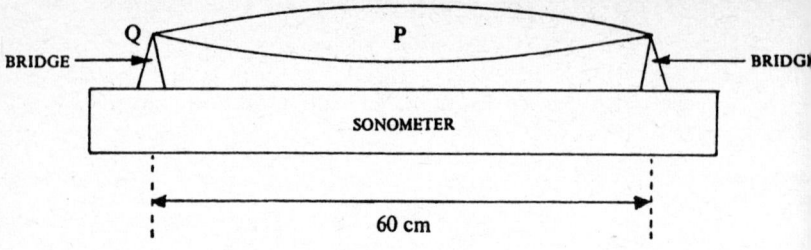

(i) What names are given to the parts of the standing wave at position P and at position Q?

(ii) Explain how the form makes the sonometer wire vibrate to produce the standing wave.

(iii) If the tuning fork emits a note of frequency 220 Hz, calculate the wavelength in air of the sound produced by the fork.

(iv) What is the wavelength of the wave produced on the wire?

(v) Describe and explain what would happen if the experiment were repeated with a fork of frequency 440 Hz. (13 marks)

Data for this Question:

speed of sound in air = 330 m/s   (A.E.B. Nov 1984)

Question 15   (answer - page 61)

PART I

(a) "Sound is a longitudinal wave motion in a material medium." Explain the meaning of

    (i)   wave

    (ii)   longitudinal

    (iii) medium

as used in the above sentence. (4 marks)

(b) (i) State the relationship between the speed of a sound wave in the air and the frequency and wavelength of that sound wave.

(ii) A signal generator produces oscillations of frequency 1.6 kHz and its output is connected to a loudspeaker. Calculate the wavelength in air of the sound emitted. (4 marks)

PART II

(c) Describe carefully, with the aid of a diagram, how you would demonstrate a standing wave. (5 marks)

(d) Sketch the mode of vibration of a given guitar string vibrating at

(i) its fundamental frequency

(ii) twice its fundamental frequency.

On each sketch indicate the positions of the nodes and the antinodes. (4 marks)

(e) What features of the vibration of a given guitar string affect

(i) the loudness of the sound produced

(ii) the quality of the sound produced? (2 marks)

(f) How is the fundamental frequency of a vibrating string affected by increasing

(i) its length

(ii) its tension? (2 marks)

Data for this Question:

The velocity of sound in air = 320 m/s. (A.E.B. Nov 1983)

Question 16 (answer - page 62)

(a) What is the essential difference between a longitudinal wave and a transverse wave? Why is the sound wave in air produced by a vibrating tuning fork longitudinal?
(4 marks)

(b)

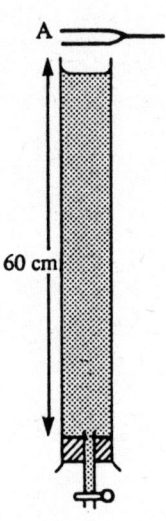

A student holds a vibrating tuning fork A, which has the number 256 engraved on it, just above a glass tube of length 60 cm containing water. The tube is allowed to slowly empty of water and the student hears a loud sound when there is 25.0 cm of water left in the tube. On repeating the experiment with tuning fork B, which has the number 512 engraved on it, he hears a loud sound on two occasions as the tube empties of water.

(i) What is the meaning of the number 256 on tuning fork A? (2 marks)

(ii) Explain why the student hears a loud sound when using fork A and state the name of this effect. (4 marks)

(iii) Use the figures given for tuning fork A to calculate a value for the speed of sound in air. Suggest a reason why this value is only approximately correct. (5 marks)

(iv) Explain why, when using tuning fork B, there were two occasions as the tube was emptying of water when a loud sound was heard, whereas with tuning fork A there was only one. (5 marks)

(U.O.L. June 1982)

Question 17 (answer - page 64)

PART I

(a) A small vertical object is placed close to a converging (convex) lens.

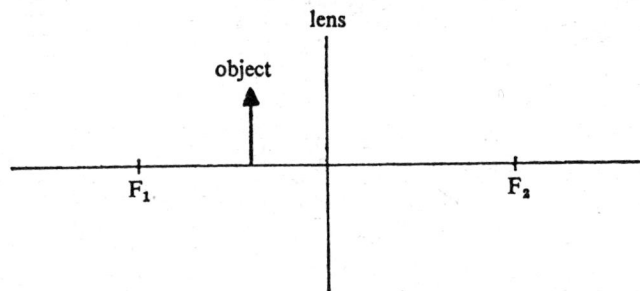

Copy the diagram in your answer book and draw a ray diagram to locate the image. Draw the image.
(4 marks)

(b) The object is moved along the axis to the left.

State what happens to the size and position of the image as the object

(i) moves nearer to F_1

(ii) moves beyond F_1. (4 marks)

PART II

(c) A student is required to make a single-lens camera to take pictures of objects between 60 cm and 300 cm from the lens of the camera. He has a light-tight box, of length approximately 11 cm, with a threaded tube containing a converging (convex) lens of focal length 10 cm. A film is positioned as shown and the shutter is normally closed.

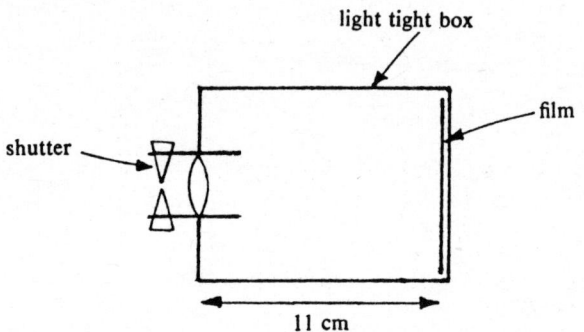

(i) How far, and in what direction, must the student move the lens from the position shown to enable the image of an object 60 cm from the lens to be focused on the film? (For solutions by a ray diagram a suitable scale is 2 cm represents 10 cm.)

(ii) Having taken the photograph of the object at 60 cm in which direction must the student move the lens in order to focus an object which is 300 cm from the lens? Justify your answer.
(8 marks)

(d) On a practical camera there is an aperture control as well as an adjustable shutter.

(i) Explain the function of the aperture control.

(ii) If you wanted to photograph a fast moving object, what adjustments would be needed? (5 marks)

(A.E.B. June 1984)

Question 18 (answer - page 65)

PART I

(a) The diagram shows a ray **AB** of monochromatic light incident at an angle of 45° to the surface of a glass block.

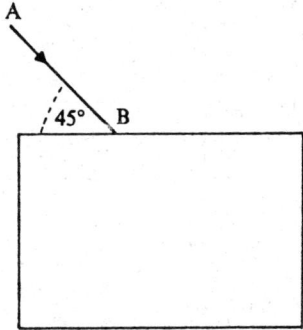

Copy the diagram in your answer book and complete it to show

(i) the approximate path of the ray through the block

(ii) the ray emerging from the opposite side.

(3 marks)

(b) Why does the ray of light take the path you have described in (a)? (5 marks)

PART II

(c) The frequency of the ray of light in Part I is 6×10^{14} Hz.
The speed of light in air is 3×10^8 m/s.
The refractive index of the glass (in air) is 1.5.

(i) What is the wavelength of the light in air?

(ii) Through what angle is the ray turned when entering the block at **B**?

(iii) What is the wavelength of the light in the glass?

(iv) What is the angle of incidence at the lower face of the glass block? (10 marks)

(d) How would each of the answers to questions (c) (i) and (c) (ii) be changed if a ray of lower frequency were substituted for the original ray? In each case state your reason. (3 marks)

(A.E.B. Nov 1984)

Question 19 (answer - page 67)

PART I

(a)

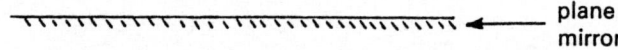 ← plane mirror

The diagram shows a letter **L** on a horizontal piece of paper on which a plane mirror is placed vertically.

(i) Copy the diagram on your paper, and draw suitable rays to locate the image of the corner of the letter **L** nearest to the mirror.

(ii) On your diagram draw the image of the letter **L**. (5 marks)

(b) Draw diagrams to show how a $45°$-$90°$-$45°$ glass prism can turn a narrow beam of light through

(i) $90°$,

(ii) $180°$.

On both your diagrams show the paths of the beams of light in the glass. (3 marks)

PART II

(c) You are given two $45°$-$90°$-$45°$ glass prisms and a cardboard tube of rectangular cross-section.

(i) Show, by diagram, how you would mount the prisms in the tube so that you could see objects behind you.

(ii) Draw two rays of light from an object, to show how you would see the image of the object.

(iii) State what is wrong with the system.

(iv) State how you would correct it. (8 marks)

(d) An object is placed on the axis of a converging (concave) mirror of focal length 200 mm. The image produced is inverted and has a magnification of 1.5. By calculation or by scale drawing on graph paper determine the position of the object. (5 marks)

(A.E.B. Nov 1982)

Question 20 (answer - page 70)

This question is about the refraction of light.

(a)

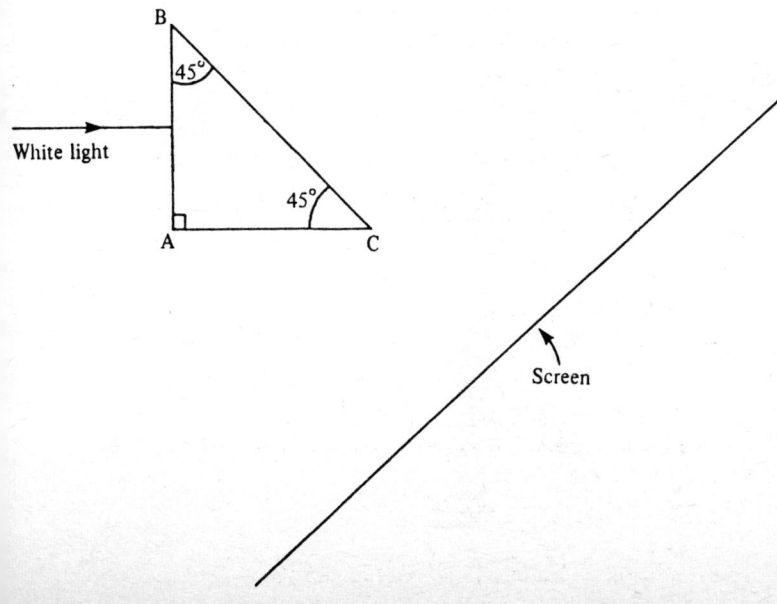

In an attempt to determine the refractive index of the glass of a prism, a student directs a narrow beam of white light perpendicularly towards the side **AB** as shown above.

(i) By describing what happens to the light as it passes into, through and out of the prism, explain why the student will find it impossible to determine a value for the refractive index of the glass using this arrangement.

(ii) Draw a diagram showing the path taken by the light into, through and out of the prism.

(iii) Give reasons why the light behaves in the way you have described in (a) (i). (6 marks)

(b) In a further experiment, the student replaces the prism shown in the diagram by one with B = $30°$ and C = $60°$.

(i) Draw a separate diagram to show how this prism produces a spectrum on the screen. State the name of the colour which is deviated the most.

(ii) Calculate the angle of emergence from the side BC of the colour for which the glass has a refractive index of 1.53. (7 marks)

(c) The speed of light in air is 3.0×10^8 m/s and the mean refractive index of glass is 1.5. Calculate the speed of light in glass.

What effect, if any, does this change in speed have on

(i) the frequency, and

(ii) the wavelength

of the light as it passes through the glass? (4 marks)

(U.O.L. June 1983)

27.

Question 21 (answer - page 71)

This question is about experimental work with a lens.

A student varies the distance of an illuminated object from a lens until a clear image is obtained on a screen. Each time the distance of the object from the lens and the linear magnification are measured.

(a) Draw a diagram of the experimental arrangement. Name and show clearly the type of lens used.
(3 marks)

(b) Describe how the magnification may be determined in this experiment. (3 marks)

(c) _____

Object distance/cm	25.0	30.0	35.0	40.0	50.0	60.0	70.0
Magnification	4.0	2.0	1.8	1.0	0.67	0.5	0.4

The table shows the results as recorded by the student. On <u>one</u> occasion, the student made an error in determining the magnification.

Plot a graph of magnification (y-axis) against object distance (x-axis).

Use the graph to determine

(i) the incorrectly determined magnification,

(ii) the correct value of the magnification for this object distance,

(iii) the focal length of the lens. (11 marks)

(U.O.L. June 1984)

Question 22 (answer - page 73)

PART I

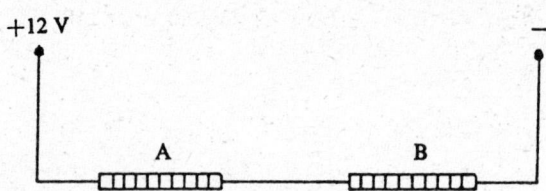

(a) When two identical heating coils **A** and **B** are connected across a 12 V d.c. supply of negligible internal resistance a current of 2 A flows through **A**. What is

 (i) the current through B

 (ii) the resistance of **A**?

 Justify your answers. (4 marks)

(b) If **A** and **B**, connected as shown in the diagram, are each immersed in equal but separate volumes of water, the temperature rise in each in the first minute is 5 K (^{0}C). Neglecting the heat capacities of the containing vessels, what would the temperature rise in the first minute if they were connected in parallel across the same supply? Explain your reasoning. (4 marks)

PART II

(c) An electric door bell is operated from the 240 V 50 Hz mains supply using a step-down transformer.

 (i) Draw a diagram of a suitable circuit.

 (ii) The transformer has 1800 turns on the primary and 90 turns on the secondary. Assuming that the transformer is ideal, what is the e.m.f. in the secondary? (6 marks)

(d) Two householders use transformers of the type described in (c). Householder X sites his transformer near the fusebox, while householder Y sites his near the bell. The long leads used by each have a total resistance of 0.5 Ω. The current in the primary coil is 0.08 A and current in the secondary coil is 1.5 A.

Householder Y claims he has a smaller voltage drop along his leads than householder X does, while X claims his method is safer.

(i) Show that Y's assertion is correct.

(ii) Explain how X's method is safer. (7 marks)

(A.E.B. Nov 1984)

Question 23 (answer - page 74)

PART I

(a) A student uses a 0 to 10 mA meter which gives its full-scale deflection when there is a potential difference of 0.1 V across it.

   (i) The student knows that he can convert the meter to give readings of higher currents by using a resistor, as shown in the diagram.

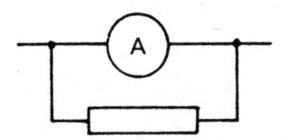

       Explain how this arrangement works. (2 marks)

   (ii) Calculate the resistance of the meter. (2 marks)

   (iii) How would the student convert the meter to obtain a full scale deflection when a potential difference of 10 V is applied across the terminals? (4 marks)

PART II

(b) The table shows the e.m.f. and short-circuit current for each of three cells.

Cell	e.m.f.	Short-circuit current
Wet Leclanche	1.5 V	3 A
Dry Leclanche	1.5 V	1.5 A
Accumulator	2.0 V	300 A

(i) Which cell has the largest internal resistance?

(ii) Calculate the internal resistance of the Wet Leclanche cell. (3 marks)

(c) A 2 Ω resistor is connected across the terminals of the Wet Leclanche cell. Assume that the internal resistance of the cell remains constant.

(i) What is the total resistance of the circuit?

(ii) What current flows through the circuit?

(iii) What is the potential difference across the terminals of the cell? (6 marks)

(d) The diagram shows a circuit suitable for charging a 12 V accumulator.

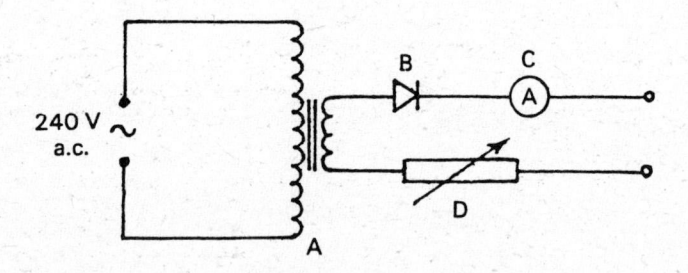

State the name and function of each of the components A, B, C, D. (4 marks)

(A.E.B. June 1983)

Question 24 (answer - page 76)

PART I

(a) Figure 1 shows two long identical wires **AB** and **CD** which may be connected to two identical d.c. sources.

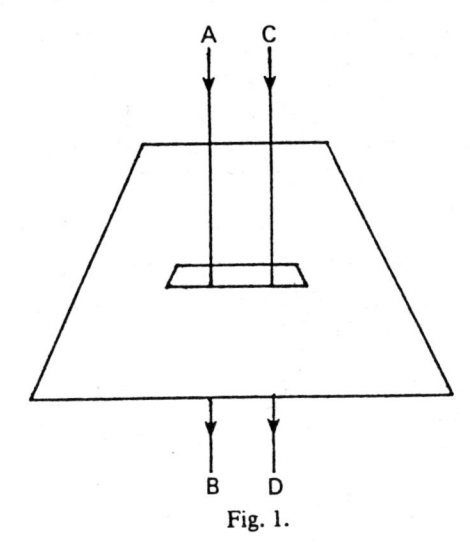

Fig. 1.

(i) Draw the magnetic field pattern due to current flowing from A to B only.

(ii) Draw the resultant magnetic field pattern due to currents flowing from A to B and from C to D.

(iii) In situation (ii), what forces, if any, will act on each of the two wires? (8 marks)

PART II

(b) A coil of wire is pivoted horizontally between the poles of a magnet as shown in Figure 2 (pivot not drawn). An electric current is passed from **E** which is positive to **F** which is negative.

Fig. 2.

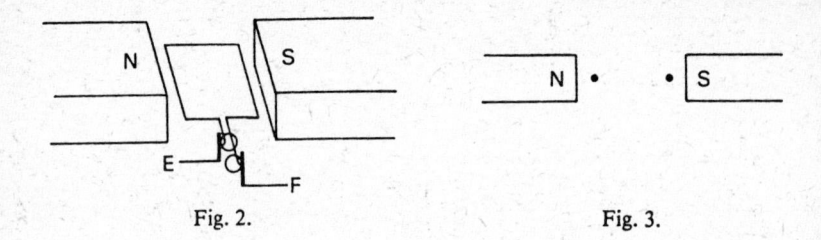

Fig. 3.

Figure 3 shows a vertical section through the centre line of the poles. Copy Figure 3 in your answer book and mark on it

(i) the forces acting on the two sides of the coil

(ii) the direction of movement of the coil. (2 marks)

(c) The apparatus shown in Figure 2 can be used to generate electricity.

(i) In which direction is the current produced if the circuit between **E** and **F** is completed and the coil is rotated clockwise from the position shown?

(ii) State the nature of and explain the variation of e.m.f. as the coil rotates through one complete revolution.

(iii) In your answer book, sketch a graph to show the variation of the e.m.f. in one complete revolution. (8 marks)

(d) Mains electricity uses a 50 Hz, 240 V r.m.s. supply.

(i) Explain the meaning of 50 Hz.

(ii) Why is the r.m.s. voltage quoted?

(iii) How is the r.m.s. voltage related to the maximum voltage? (3 marks)

(A.E.B. June 1983)

Question 25 (answer - page 78)

PART I

(a) Draw a circuit diagram to show how three identical lamps, A, B and C, all required to work at normal brightness, can be connected to a supply so that A and B are together controlled by one switch and C is controlled by a second switch. (4 marks)

(b) If a fuse of average resistance 0.01 Ω takes 5 s to melt when a current of 6 A passes through it, calculate the energy needed to melt the fuse.

PART II

(c) A small electric heater consists of two elements of resistance 60 Ω and 120 Ω respectively.

The heater operates on 240 V mains and the two elements can be connected to give four different heat outputs.

(i) Draw the arrangements of the elements for the four outputs. What is the effective resistance of each of the arrangements?

(ii) What is the current taken by the heater at maximum output?

(iii) The heater costs 48p to run over a period of 10 hours when the price of electricity is 5p per kWh. What is the average current during the 10 hour period of use? (13 marks)

(A.E.B. Nov 1983)

Question 26 (answer - page 79.)

This question is about the experimental determination of resistance and the use of a temperature-dependent resistor.

(a) (i) Draw a circuit diagram of an ammeter/voltmeter method for investigating how the resistance of the filament of a lamp varies with the current.

    (ii) Name the component in the circuit which enables the current to be varied and explain how this is achieved.

    (iii) List the readings taken and show how the resistance is calculated.

    (iv) Assuming that the filament is made of the metal tungsten, how would you expect the resistance to vary with the current? Why does the resistance vary in this way? (11 marks)

(b)

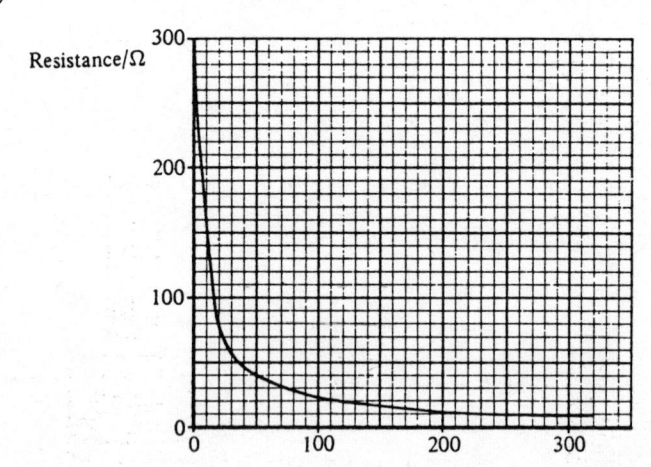

[Graph: Resistance/Ω (y-axis, 0 to 300) vs Temperature/°C (x-axis, 0 to 300), showing a curve that decreases rapidly from about 300 Ω near 0 °C, levelling off close to 0 Ω by about 100 °C.]

The above graph shows how the electrical resistance of a temperature-dependent resistor (thermistor) varies with temperature.

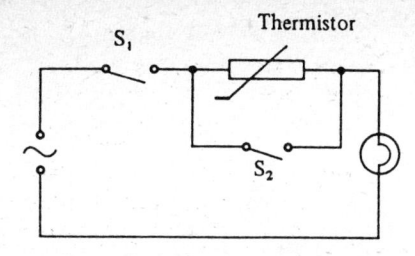

35.

The circuit above shows a thermistor being used to prevent a large initial current in a delicate lamp filament when the switch S_1 is closed and yet allowing the current to increase to its normal value when the filament has warmed up.

(i) Describe how the thermistor prevents the large initial current.

(ii) A short time after S_1 is closed, the switch S_2 is closed. Explain why S_2 is included in the circuit. (6 marks)

(U.O.L. June 1984)

Question 27 (answer - page 80)

(a)

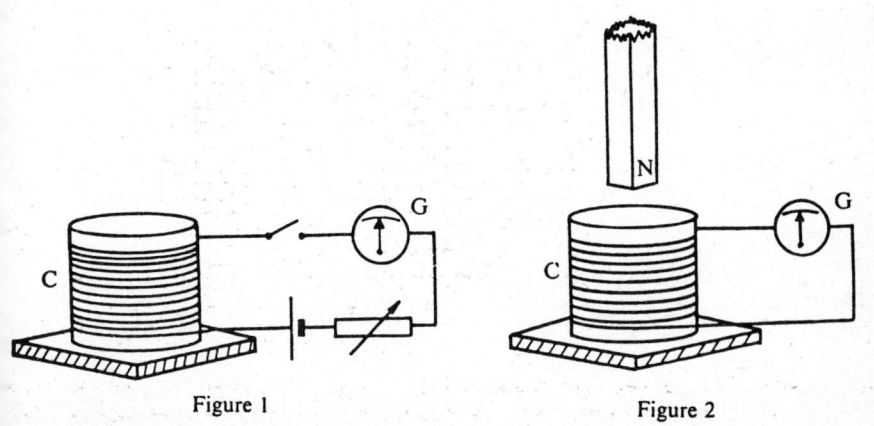

Figure 1

Figure 2

Figure 1 shows a hollow coil **C** of about 200 turns mounted on a wooden base. **G** is a centre-zero galvanometer. When the switch is closed the galvanometer needle is deflected to the right.

Suppose the coil is connected to the galvanometer as in Figure 2 and the N-pole of a long bar magnet is

(i)   lowered quickly into the coil,

(ii) left at rest on the base of the coil,

(iii) moved quickly in and out of the coil about twice per second.

Describe and explain the resulting movements of the galvanometer needle in each case.   (9 marks)

(b) The diagram below shows a transformer which has a primary coil of 400 turns of insulated copper wire, and a secondary coil of 5 turns of thick walled copper tubing. The primary coil is connected to the 240-V a.c. mains, and a nail is connected across the secondary terminals. When the switch S is closed the nail glows and then melts.

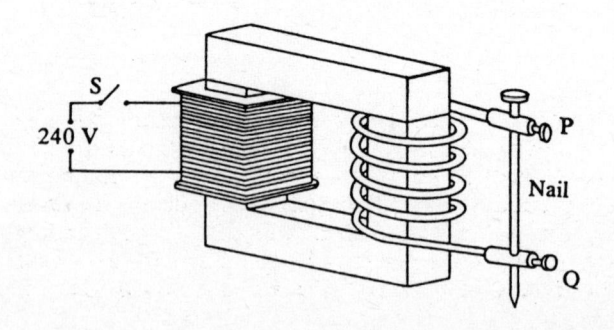

37.

Explain the action of the transformer, making it clear why a voltage is produced across the ends of the nail. Calculate a value for this voltage.
(5 marks)

Explain

(i) why the nail melts, (2 marks)

(ii) why the transformer is more efficient if the core is laminated, (2 marks)

(iii) why, if the top bar of the core is removed before the switch is closed, the nail remains quite cool. (2 marks)

(U.O.L. Jan 1981)

Question 28 (answer - page 82)

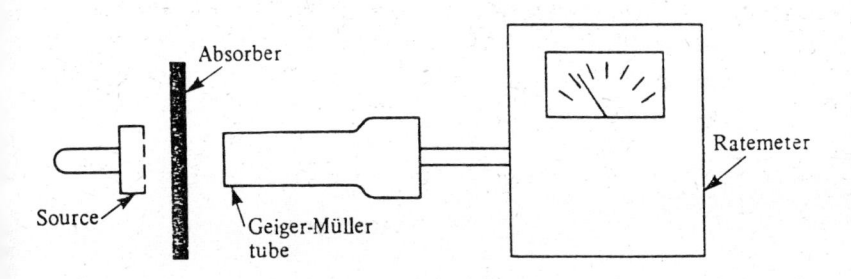

A weak radioactive source was thought to emit β-radiation. In an attempt to confirm this a student arranged a Geiger-Müller tube, connected to a ratemeter, close to the source and then placed sheets of different materials as absorbers between the source and the tube. Three readings of the ratemeter were taken at 10-second intervals for each sheet and the results were tabulated as follows:

Absorber material	Ratemeter reading counts per minute		
	1	2	3
Air	120	110	130
Paper	100	120	110
Cardboard	130	130	100
Aluminium (0.5 mm thick)	110	120	110
Aluminium (5 mm thick)	50	60	40
Lead (5 mm thick)	40	50	50
Lead (50 mm thick)	50	40	50

(a) Give reasons for confirming that the source does not emit α-rays or γ-rays but must emit β-radiation. (4 marks)

(b) Suggest another test, not involving absorption, you could perform to further confirm that the source emitted β-radiation. (4 marks)

(c) As only β-radiation was emitted, account for the ratemeter readings for the 50-mm thick lead absorber. (3 marks)

(d) Why were identical readings not obtained each time for the same absorber? (3 marks)

(e) If the source had been emitting either α-radiation or γ-radiation a different set of readings would have been obtained. Copy the table below for an α-source and a γ-source and complete the table by inserting suitable estimated values you would expect to obtain for each absorber.

Absorber material	Ratemeter reading counts per minute	
	α-source	γ-source
Air	120	120
Paper		
Cardboard		
Aluminium (0.5 mm thick)		
Aluminium (5 mm thick)		
Lead (5 mm thick)		
Lead (50 mm thick)		

(6 marks)

(U.O.L. June 1981)

Question 29 (answer - page 83)

(a)

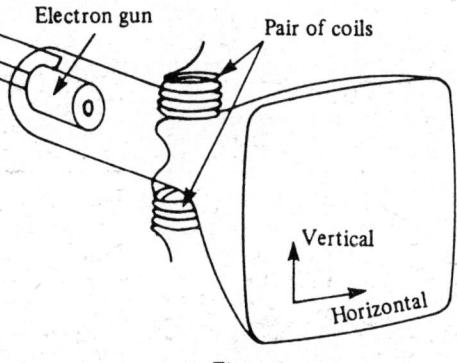

Figure 1

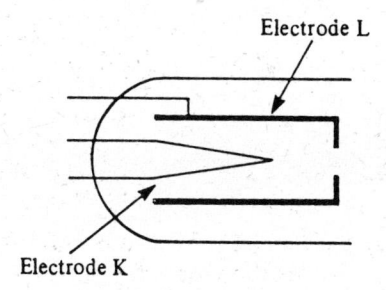

Figure 2

The cathode ray tube of a black and white television is shown in figure 1. The 'electron gun', which is the assembly responsible for the generation of cathode rays, is shown in figure 2. Describe how the electron gun works. **(6 marks)**

What is the purpose of the pair of coils shown in figure 1 and how is this purpose achieved?
(4 marks)

(b) A weak radioactive sample was fixed in a holder and monitored every day for five days. The monitoring was achieved by placing a Geiger-Muller tube connected to a ratemeter in contact with the sample. Three readings of the ratemeter were taken at 5-minute intervals at the same hour every day. The following results, corrected for background radiation, were obtained.

Day	Mon	Tues	Wed	Thur	Fri
Corrected ratemeter readings / count/min	595	380	245	160	97
	605	378	250	140	103
	600	382	255	150	100

Construct a table showing the average ratemeter reading for each day of the week. (2 marks)

Draw a graph of these readings against time, using the graph paper supplied. (5 marks)

Determine the half-life of the radioactive sample. (3 marks)

(U.O.L. June 1982)

Question 30   (answer - page 86)

(a) Suppose you are supplied with a negatively charged rod and two identical and uncharged metal spheres, A and B, on insulating stands.

   (i) How, using the negatively charged rod, would you charge the two spheres equally so that A is negatively charged and B is positively charged? (4 marks)

   (ii) How would you show experimentally, using any additional apparatus you may require, that this had, in fact, occurred? (6 marks)

   (iii) Account for the production of the two equal charges in terms of the movement of charged particles. (4 marks)

(b) A p.d. of 2 V is applied across the ends of a resistor of value 5 MΩ (5×10^6 Ω). Calculate

    (i) the current in the resistor,

    (ii) the charge which flows through the resistor in 1 s, and

    (iii) the number of electrons which flow through the resistor in 1 s.

(The charge carried by one electron is 1.6×10^{-19} C).

(6 marks)

(U.O.L. Jan 1981)

Question 1

PART I

(a) (i) Length increases.
    (ii) External diameter increases.
    (iii) Internal diameter increases.

(b) Thermal conduction can be demonstrated in the laboratory by coating one end of such a pipe with a thin layer of melted candle wax, allowing it to cool, and then applying heat by means of a bunsen flame to the non-waxed end. The conducted heat causes the wax to melt.

PART II

(c) Volume of pipe $= \dfrac{\text{mass}}{\text{density}} = \dfrac{22}{11000} \times 1000000 = 2000 \text{cm}^3$

but volume = length x area = L x 50 = 50 L cm^3

∴ 50L = 2000

∴ L = 40 cm   <u>Length of pipe = 40 cm</u>

(d) (i) <u>Change in length:</u>- $l = L \propto \theta$ {L = original length
      $= 40 \times 3 \times 10^{-5} \times (490-290)$ $\propto$ = linear expansivity
      = <u>0.24 cm</u> θ = temp. change}

   (ii) <u>Change in thickness:</u>- $t = T \propto \theta$ {T = original thickness}
      $= (10-6) \times 3 \times 10^{-5} \times (490-290)$
      = <u>0.024 cm</u>

   (iii) <u>Energy</u> = $mc\theta$ {m = mass, c = sp. heat capacity}
      $= 22 \times 125 \times (490-290)$
      = <u>5.5 K-joules</u>

   (iv) <u>Energy</u> = mL   {L = sp. latent heat of fusion}
      ∴ $5.5 \times 10^6 = 22 \times L$, ∴ $L = 2.5 \times 10^5$ J/kg
      <u>Specific Latent heat of fusion of lead</u>
      <u>= 2.5×10^5 J/kg</u>

Question 2

PART I

(a) **Specific heat capacity** of a substance is the amount of heat required to raise the temperature of unit mass (1 kg) of the substance by $1^0 C$.

Specific latent heat of a substance is the amount of heat required to melt unit mass at the constant temperature of its melting point:- (Fusion) or the amount of heat required to boil unit mass at the constant temperature of its boiling point:- (Vapourization).

(b) Heat is supplied at the same, constant, rate in each case. The rate of increase in temperature of the glycerol being **twice** that of the water indicates that the specific heat capacity of glycerol is **half** that of water. As temperature continues to **rise** until boiling point is reached, we conclude that boiling point of water is 373 K ($100^0 C$) and that of glycerol is 563 K ($300^0 C$).

PART II

(c) (i) **Potential difference:-** "V = IR"

$$\therefore V = 4 \times 6 = \underline{24 \text{ volts}}$$

(ii) **Power output of coil:-** "W = $I^2 R$"

$$\therefore W = 4^2 \times 6 = \underline{96 \text{ watts}}$$

(iii) **Energy:-** "J = Wt"

$$\therefore J = 96 \times 60 = 5760 \text{ joules}$$

Energy supplied in 1 minute = $\underline{5760 \text{ joules}}$

(iv) **Heat gained by water:-** "Q = mct"

$$\therefore Q = 0.25 \times 4200 \times (25 - 20) = 5250 \text{ joules}$$

$$\text{Average loss of energy} = \frac{(5760 - 5250)}{60}$$

$$= 8.5 \text{ joules/sec.}$$

$$= \underline{8.5 \text{ watts}}$$

(d) As the temperature of water increases, so does that of the immersed coil. The resistance of the coil increases as its temperature increases and, as the potential difference remains constant, the current through the coil will decrease.

Question 3

PART I

(a) Two factors affecting boiling point are:-

   (i) <u>Pressure</u>   If external pressure is increased, the boiling point increases, and vice versa.

   (ii) <u>Impurities</u>   If impurities are added the boiling point may change. For example, the addition of salt will cause the boiling point to be increased.

(b) Evaporation takes place at any temperature and only at the surface. Boiling occurs at a definite temperature and within the liquid.

(c) The mercury in the right hand tube is lower than that in the left hand tube because the vapour exerts a 'vapour pressure' on the mercury.

PART II

(d) (i) The cooling tubes are in the roof because, by convection, the warmest air within the store will be in contact with the ceiling. As this air is cooled, it is replaced by the now-warmer air beneath it and the store gets progressively cooler until the desired store temperature is attained.

   (ii) The refrigerating machine is outside the store because it contains the compressor which converts the circulating coolant from the vapour state to the liquid state, which results in latent heat being released to the surrounding air. If the machine were inside, the store room air would be re-heated and the temperature would not decrease.

(iii) When the store room temperature falls to that of the liquid coolant, no further heat transfer is possible.

(e) (i) When the block is placed in the water, it gains heat from the water which is at a higher temperature. As the water is at $0°C$ it will not experience a fall in temperature until all of it has changed to ice. The heat transfer is therefore due to the water in the immediate vicinity of the block giving up latent heat as a layer of ice is formed on the block.

(ii) The temperature will be $0°C$.

(iii) Heat gained by copper block

= mass x sp. ht. capacity x temp. rise

= 0.68 x 400 x 50 joules

Heat lost by water changing to ice

= mass of ice x sp. latent heat of ice

= m x 340000 joules

"Heat lost = Heat gained"

∴ m x 340000 = 0.68 x 400 x 50

∴ m = 0.04 kg

Mass of ice formed = 40 g

Question 4

PART I

(a) The Saturated Vapour Pressure of a liquid is the pressure exerted on the liquid by its vapour when the liquid and vapour are in contact with each other within a closed container and a state of dynamic equilibrium has been reached (vapour is saturated). S.V.P. increases if temperature increases, and vice versa.

(b) (i) If, at a constant temperature, the vapour is removed by a vacuum pump - the pressure exerted on liquid decreases and there is dynamic imbalance. More molecules leave the liquid and boiling takes place.

(ii) If the pump is switched off, boiling will continue until the pressure builds up to the S.V.P. of the liquid and there is, again, dynamic equilibrium.

(iii) The temperature of the water will fall as it boils, because the heat required to vapourize the water is not being provided by an external source but has to be provided by the water.

PART II

(c) From the graph:-

(i) Temperature corresponding to S.V.P. of 100 k Pa
$= \underline{100^0 C}$

(ii) For a boiling point of 200^0, pressure = $\underline{1550 \text{ k Pa.}}$

(iii) The temperature corresponding to a pressure of 1200 k Pa = $\underline{188^0 C}$.

(d) (i) Heat given up by condensing steam

= mass x sp. latent heat of vapourization

= 0.1 x 2.26 = 226000 J/minute

Heat given up by condensed steam cooling through

(100 - 70):- 30 K = mass x sp. heat capacity x 30

= 0.1 x 4200 x 30 = 12600 J/min

∴ Total energy evolved per minute

= 226000 + 12600

= $\underline{238.6 \text{ kJ}}$

(ii) Assuming that all the energy evolved is gained by the cooling water, we have:-

Heat gained by water per minute

= 1 x 4200 x θ (where θ = rise in temp.)

(continued.....)

$\therefore$ 4200 θ = 238.6 × 1000

$\therefore$ θ = 56.8 K

$\therefore$ Leaving temperature of cooling water
= (5 + 56.8)°C = 61.8°C

Question 5

PART I

(a)

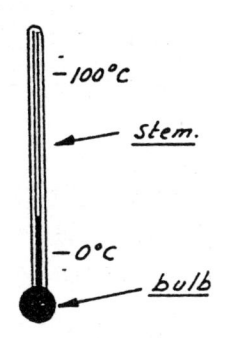

```
        |—100°C
        ← stem.
        
        —0°C
        ← bulb
```

Lower fixed point:- melting point of ice at normal atmospheric pressure:- 0°C.

Upper fixed point:- boiling point of water at normal atmospheric pressure:- 100°C.

Stem consists of a, relatively, thick glass tube with a fine bore allowing large movement of mercury for small changes of temperature. Bulb glass is, relatively, thin.

(b) (i) Dissolved solids increase the boiling point and decrease the freezing point of water.

(ii) Increase in atmospheric pressure increases the boiling point and decreases the freezing point of water. Decrease of pressure has the opposite effect.

PART II

(c)

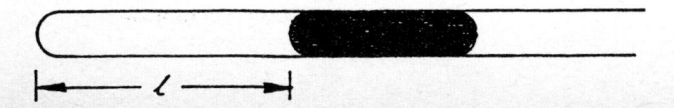

|←— ℓ —→|

If pressure remains constant, volume of trapped air is proportional to the temperature. If the cross-sectional area of the tube is considered to remain constant within the range of temperatures being considered, then temperature is proportional to l.

Difference in temp. between lower and upper fixed points = $100°C$.

∴ (178 - 128) = 50 mm represents a change in temp. of $100°C$.

∴ 1 mm represents a change in temp. of $2°C$.

(i) Room temp. = $0° + 2(136-128) = 16°C$

(ii) At $-10°C$, length = $(128 - \frac{10}{2})$mm = 123 mm

(d)

```
┌─────────────────────────┐
│      EVAPORATOR         │
│    Expansion            │
│      valve              │
│           CONDENSER     │
│  Vapour                 │
│         COMPRESSOR      │
│   low        high       │
└─────────────────────────┘
```

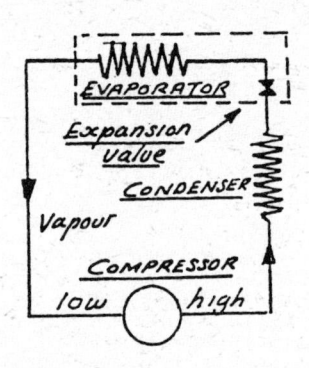

In the normal - inoperative - condition, the coolant is in its liquid state. With the compressor operating, the pressure on the evaporator side is reduced causing a reduction in boiling point and the consequent vapourization of the liquid within the evaporator. The latent heat necessary for the vapourization is taken from the contents of the refrigerator. The pressure on the condenser side is increased causing an increase in boiling point and the consequent liquifaction of the vapour within the condenser. The latent heat is given up to the atmosphere.

(e) The cooling unit is situated near the top because the warmest air is in this region - because of convection of air within the refrigerator.

Question 6

(a) The liquid will tend to be warmer at the top of the container because the warmer, less-dense liquid is subjected to an upthrust by the surrounding cooler, more-dense liquid which replaces it and sets up a convection current within the liquid.

(b) (i) Glass fibre would reduce heat losses by conduction and radiation from the container to its surroundings.

(ii) A lid would reduce loss of heat by convection.

(c) The temperature will stop rising because the rate of heat loss increases as temperature increases. When the energy loss is equal to the amount of energy being provided, temperature ceases to rise.

(d) This is done to compensate for loss of heat to the surroundings whilst the temperature of the liquid is above room temperature. The heat gained from the surroundings before room temperature is attained is equal to the loss of heat to the surroundings after it has been attained.

Neglecting losses, energy input = energy gained by liquid

$\therefore$ per second:- $36 = mc\theta$ m = mass
$\therefore \quad\quad\quad\quad 36 = 0.5 \times 4200 \times \theta$ c = sp. ht. capacity

$$\frac{36}{0.5 \times 4200} = \theta$$

θ = temp. rise

$\therefore$ Rise in temp. per minute = $\dfrac{36 \times 60}{0.5 \times 4200}$ = 1.03 K

If allowance is made for the container, the rate of temperature rise would be less because some of the input energy would be utilized in raising the temperature of the container as well as that of the liquid it contains.

Question 7

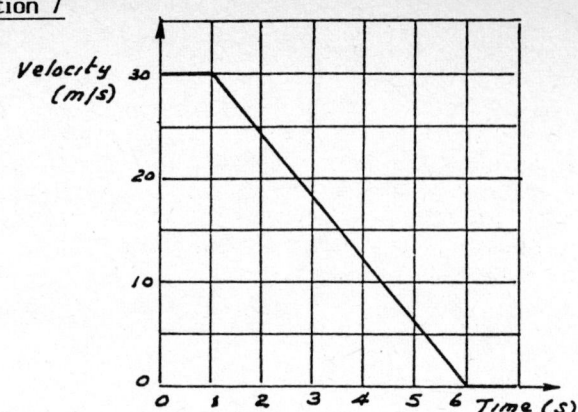

[Graph: Velocity (m/s) vs Time (s). Line starts at 30 m/s, horizontal from t=0 to t=1, then decreases linearly to 0 at t=6.]

(a) The distance travelled by the car before it stops is given by the area under the graph from 0 to 6 secs.

Area = $(30 \times 1) + \frac{1}{2}(30 \times 5)$ = 105 sq. units

Car travels 105 m and, therefore, does not collide.

(b) Retardation = $\dfrac{30 - 0}{6}$ = 5 m/s^2

Braking force = mass × retardation = 1500 × 5 = 7500 N

(c) As the car slows down, the velocity of driver remains at 30 m/s. The driver is moving faster than the car.

(d) Kinetic Energy = $\frac{1}{2}mv^2$ = $\frac{1}{2} \times 1500 \times 30^2$ = 675 kJ

(e) If the car had been travelling at 15 m/s, its kinetic energy would have been 1/4 of that at 30 m/s. If the braking distance is dependent only upon braking force, then it would be 1/4 of the distance at 30 m/s. However, wind resistance is also a factor and this is greater at higher speeds and is, thus, of less significance at 15 m/s. Consequently, the braking distance would have been more than 1/4 of that at 30 m/s.

Question 8

PART I

(a) (i) The piece of paper and the apple are each subject to two forces:- gravitational force and air resistance. In the case of the paper, the ratio:- gravitational force : air resistance is less than in the case of the apple. Consequently, the paper takes longer to reach the ground.

    (ii) On the moon, only the moon's gravitational force is acting (with no air to resist motion). They both reach the surface at the same time.

(b) With usual convention

    u = 0   "v = u + at"

    v = 5 m/s $\therefore$ 5 = 0 + (a × 3)

    t = 3 s $\therefore$ a = 5/3 m/s^2

    <u>Acceleration due to gravity on moon = 5/3 m/s^2</u>

PART II

(c) (i) <u>Loss of P.E.</u> = "mgh" = 100 × 10 × 320

                            = <u>320000 J</u>

    (ii) <u>Gain in K.E.</u> = Loss in P.E.

                     = <u>320000 J</u>

    (iii) "K.E. = 1/2 mv^2"

        $\therefore$ 320000 = $\frac{1}{2}$ × 100 × v^2

        $\therefore$ v^2 = 6400

        $\therefore$ v = 80 m/s

    <u>Vertical component of speed = 80 m/s</u>

    (horizontal component remains constant)

(continued.....)

(d) (i)

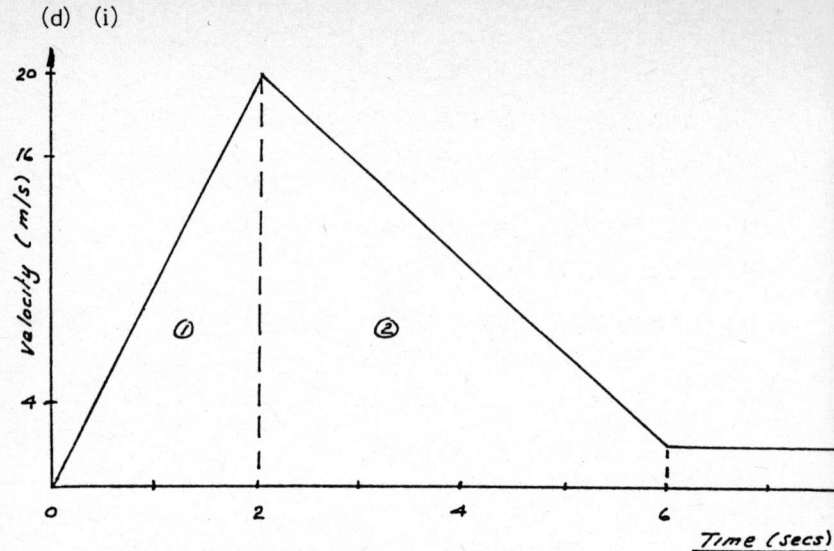

[Graph: velocity (m/s) vs Time (secs). Triangle rising from (0,0) to (2,20), falling to (6, ~2), then constant. Areas labeled ① (0-2) and ② (2-6).]

(ii) Max velocity (after 2-secs)

"$v = u + at$" ∴ $v = 0 + (10 \times 2) = 20$ m/s

Distance fallen (0-2 secs) - represented by

$= \frac{1}{2} \times 2 \times 20 = 20$ m

Distance fallen (2-6 secs) - represented by area (2)

$= \frac{1}{2}(20 + 2)(6 - 2) = 44$ m

Distance fallen (first 6 secs) = 20 + 44 = <u>64 m</u>

(iii) To reach ground, further falling distance =

320 - 64 = 256 m

At constant vertical speed of 2 m/s, it will take a further $\frac{256}{2} = 128$ secs.

Total time to reach ground = 6 + 128 = <u>134 secs</u>

Question 9

PART I

(a) (i) Potential energy (relative to sea level) = mgh joules

    (ii) Kinetic energy = $\frac{1}{2}mv^2$ joules

    (iii) Electrical energy = I^2R joules

    (iv) Heat energy = $ms\theta$ joules

(b) (i) Work done against gravity = mgh = w × h =
        2000 × 0.8 = 1600 joules

    (ii) If P newtons is equal to the effort required to move the load up the ramp, then the work done by the effort = P × 4 = 4P joules. If friction is neglected, work done by effort = work done on the load:-

        4P = 1600

        ∴ P = 400 N

    ∴ Effort is only one-fifth of the load.

    (iii) Work done by effort must be greater than that done against gravity because, in addition, work must be done to overcome the frictional forces between the load and ramp. Therefore, in this case, the effort will be greater than the 400 N obtained in part (ii).

PART II

(c) (i) Potential energy gained = mgh

$$= \frac{50}{1000} \times 10 \times 4.05$$

$$= 2.025 \text{ joules}$$

    (ii) Kinetic energy of the stone as it leaves the catapult is equal to the gain in potential energy.

        ∴ $\frac{1}{2}mv^2 = mgh$

        ∴ $v^2 = 2gh$

$\therefore$ $v^2 = 2 \times 10 \times 4.05$

$\therefore$ $v^2 = 81$

$\therefore$ $v = 9\ ms^{-1}$

Speed of the stone = $9\ ms^{-1}$

(d) (i) Potential energy gained by cable car:- A to B

= mgh = $4 \times 10 \times 2$ = 80 joules.

(ii) Work output = 80 joules

$$\text{Power output} = \frac{\text{work output}}{\text{time}} = \frac{80}{20}$$

$$= 4\ \text{joules/sec}$$

$$= 4\ \text{watts}$$

(iii) Efficiency = $\dfrac{\text{Output Power}}{\text{Input Power}}$

$\therefore$ 80% = $\dfrac{4}{\text{Input Power}}$

$\therefore$ Input Power = $\dfrac{4}{80\%}$ = 5 watts

Question 10

(a) Maximum speed of the train = $24\ ms^{-1}$

(b) Acceleration during the first two minutes

$$= \frac{24 - 0}{2 \times 60} = 0.2\ ms^{-2}$$

(c) The train is slowing down from 7 to 10 mins.

= 3 minutes duration

(d) Distance between the stations is represented by the total area under the graph

= $\frac{1}{2}((2-0) \times 60 \times 24) + ((7-2) \times 60 \times 24)$

    + $\frac{1}{2}((10-7) \times 60 \times 24)$

= 10800 m

(e) Average speed = $\dfrac{\text{total distance}}{\text{total time}}$ = $\dfrac{10800}{10 \times 60}$

$\qquad\qquad\qquad\qquad\qquad = 18 \text{ ms}^{-1}$

During the first two minutes, the pull is required to overcome the resistive and acceleration forces. During the next five minutes there is no acceleration and only the resistive forces have to be overcome.

During the constant speed part of the journey (2-7 mins), as there is no acceleration and the track is level (no gravitational force), the pull of the engine is equal to the resistive force.

Question 11

(a) The density of a substance is its mass per unit volume and may be quoted in kg-m^{-3}. Its relative density is its density relative to that of water and is equal to:-

$\dfrac{\text{mass of the substance}}{\text{mass of an equal volume of water}}$

There are no units.

(b) (i) The bottle has a tapered glass stopper having a fine hole through its centre allowing excess liquid to escape as the stopper is placed in the bottle as far as possible. Provided that overflow occurs, amount of liquid is a maximum.

(ii) If the bottle is closely held, more liquid may escape due to expansion caused by heat from the hand.

(iii) The following readings are taken and are used to determine the relative density as shown.

Weight of bottle:- empty $\qquad\quad$ = w1

Weight of bottle:- full of water = w2

Weight of bottle:- full of liquid $\ $ = w3

Relative density of the liquid = $\dfrac{w3 - w1}{w2 - w1}$

(c) Upthrust on B = weight of liquid displaced (Archimedes Principle)

Upthrust = weight of B + tension in chain
= (200 × 10) + 500
= 2500 N

∴ Mass of liquid displaced = $\frac{2500}{10}$ = 250 kg

If the body is moved into salty water, the same volume of liquid will be displaced and, because the salty water will have a greater density than fresh water, the mass (and thus weight) of liquid displaced will increase. The upthrust will be greater and the tension will <u>increase</u>.

Question 12

PART I

(a) When a body is partially or completely immersed in a fluid, the upthrust on it is equal to the weight of the displaced fluid (Archimede's Principle).

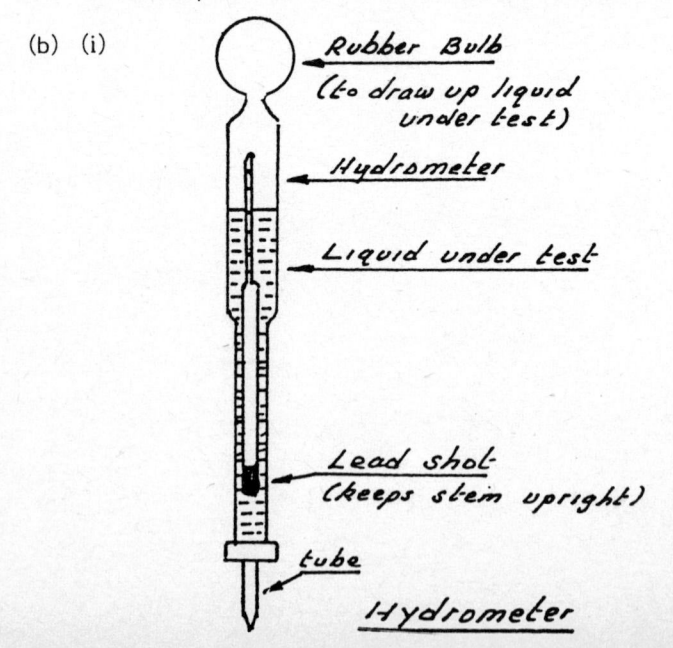

(b) (i)

Rubber Bulb (to draw up liquid under test)

Hydrometer

Liquid under test

Lead shot (keeps stem upright)

tube

<u>Hydrometer</u>

(ii) A hydrometer measures the relative density of a liquid.

(c) Pressure is the force per unit area acting on a body.

Pressure = $\dfrac{\text{Force}}{\text{Area}}$ Nm^{-2} (Pa)

Pressure due to a column of liquid:-

Pressure = hpg Pa

where h = column height (m)

p = density of liquid (kg.m^{-3})

g = acceleration due to gravity (ms^{-2})

PART II

(d) (i)

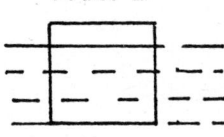

Weight of block = mg - volume x density x g

= 0.2 x 600 x 10 = 1200 N

Upthrust = weight of liquid displaced

= V x 800 x 10

= 8000 V. N

As the block is floating, 8000 V = 1200

∴ V = 0.15 m^3

Fraction of wood immersed = $\dfrac{0.15}{0.2}$ = 0.75 = $\dfrac{3}{4}$

(ii)

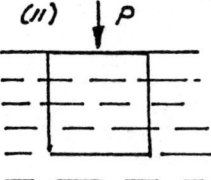

Let P = additional force

Total Force down = (P + 1200) N

Force up = 0.2 x 800 x 10 = 1600 N

∴ P + 1200 = 1600

∴ P = <u>400 N = Applied force</u>

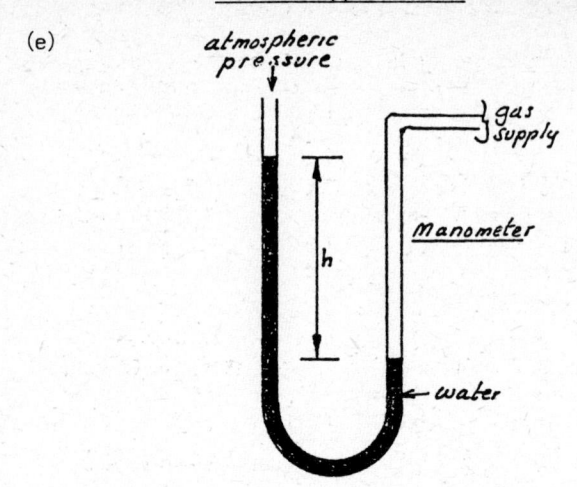

(e)

atmospheric pressure ↓

gas supply

Manometer

h

water

The manometer consists of a U-tube containing water (mercury if high pressures are being determined). When connected to the gas supply, there is a difference in water levels, as shown. The difference - h metres - is measured.

The pressure (above atmospheric pressure) of the gas supply is determined

∴ P = hpg (Pa)

(P = density of the liquid)

<u>Pressure of gas = hpg + atmospheric pressure</u>

Question 13

(a) A solid maintains its definite shape by means of intermolecular forces i.e. molecules attracting or repelling other forces in their vicinity and, consequently, vibrating about fixed positions. Increase in temperature results in the molecules obtaining more energy and, thus, more velocity and vibrations of greater amplitude. In liquids, the molecular movement is increased and the regular pattern is not possible to maintain. However, intermolecular forces are still present and are sufficient to maintain the liquid's shape within the confines of the containing vessel. In a gas, the relatively higher temperature results in even greater molecular speeds. Molecules break free from each other to move throughout the container which prevents their escape into its surroundings.

(i) Energy must be removed to reduce the molecular activity and thus provide conditions under which the molecular structure necessary for a solid is possible.

(ii) In evaporation the faster moving, higher energy surface molecules escape, thus reducing the average energy, and temperature, of the liquid as a whole.

When volume is reduced, at constant temperature, the number of molecular collisions with the container walls remains the same as does the total force due to these collisions. For same force and less area the resulting pressure must be increased.

(b) <u>Energy loss required</u>:- Water from $20°C$ to $0°C$

:- "$mc\theta$" = $0.5 \times 4200 \times 20$

= 42000 J

Water to ice at $0°C$

:- "mL" = 0.5×336000

= 168000 J

Required Heat Loss = $\dfrac{(42000 + 168000)}{2 \times 60 \times 60}$ J/s

= 29.2 J/s

∴ <u>NOT possible with heat loss of 20 J/s</u>

Question 14

PART I

(a) (i) Amplitude is equal to length of arc AB or, alternatively, α.

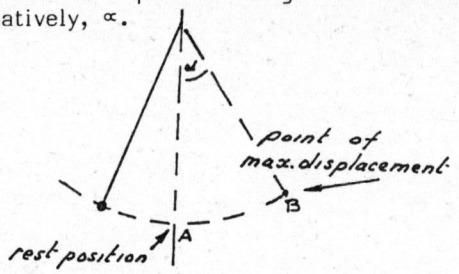

[Diagram: pendulum swing showing rest position at A, point of max. displacement at B, with arc AB]

(ii) Period of oscillation is the time taken for one complete oscillation.

(iii) Frequency is the number of complete oscillations per second.

(iv)

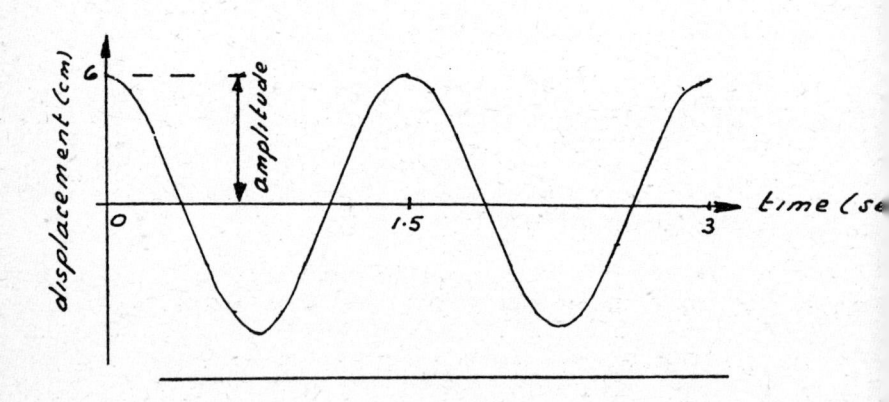

[Graph: displacement (cm) vs time (sec), sinusoidal wave with amplitude 6 cm, zero crossings at 0, 1.5, and 3, labelled "amplitude"]

PART II

(b) (i) P is an <u>antinode</u>, Q is a <u>node</u>.

(ii) The stem of the vibrating fork is held against one of the bridges and the length or tension of the wire is adjusted until the natural frequency equals that of the fork. The wire then vibrates in sympathy (resonance).

(iii) velocity = frequency x wavelength

$\therefore$ wavelength = $\frac{330}{220}$ = 1.5 m

(iv) The length of 60 cm is equal to half the wavelength.

$\therefore$ wavelength = 120 cm

= 1.2 m

(v) With a fork of frequency 440 Hz, the wavelength in air of the sound would be $\frac{1.5}{2}$ = 0.75 m.

With the wire tension kept the same, the new resonance length of the wire would be half the previous length, i.e. 30 cm.

Question 15

PART I

(a) (i) A 'wave' is a disturbance - usually periodic - which travels through a medium remaining unchanged in type.

(ii) 'Longitudinal':- the wave particles travel in the same direction as the wave.

(iii) 'Medium':- the material (solid, liquid or gas) which is necessary for the wave to travel.

(b) (i) Speed of sound in air = wave frequency x wavelength in air

(ii) $V = f\lambda$ $\therefore$ wavelength = $\lambda = \frac{V}{f} = \frac{320}{1.6 \times 10^3}$ = 0.2 m

PART II

(c)

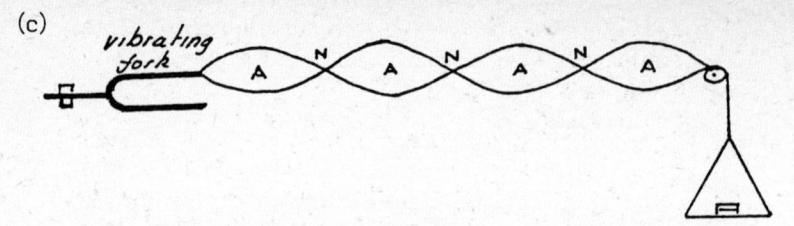

vibrating fork

One end of a string is passed over a pulley and attached to a weight to keep it taut when the other end is attached to one prong of a vibrating tuning fork. The incident and reflected waves form a standing (stationary) wave as shown. N and A indicate nodes and antinodes.

(d)

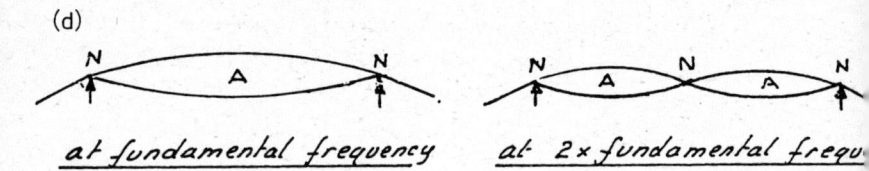

at fundamental frequency   *at 2 × fundamental frequ[ency]*

(e) (i) Loudness depends upon magnitude of vibration.

(ii) Quality of sound depends upon the presence of harmonics (frequencies other than fundamental).

(f) (i) If length is increased, fundamental frequency decreases.

(ii) If tension is increased, fundamental frequency increases.

$$f \propto \frac{1}{\text{length}}, \quad f \propto \sqrt{\text{tension}}$$

Question 16

(a) A <u>longitudinal</u> wave is one in which the particles of the medium vibrate in the direction of propogation of the wave, whilst a <u>transverse</u> wave is one in which they vibrate at right angles to the direction of propogation. A vibrating tuning fork causes compressions and rarefactions of the air which does not move as a whole towards the observer. The resulting wave motion is in the direction of propogation:- longitudinal.

(b) (i) 256 indicates the frequency of vibration:- 256 Hz.

    (ii) As the tube empties of water, the air column increases in length. When the level is reached such that the natural frequency of the air column is matched by that of the fork, the wave set up reinforces the sound of the fork and a large sound is heard. The effect is called <u>resonance</u>.

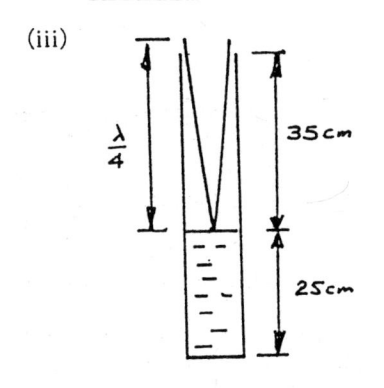

(iii) The first position of resonance is when air column is equal to $\lambda/4$

$\lambda/4 = 35$ cm

"$V = f\lambda$"

$\therefore V = 256 \times (35 \times 4)$

$\qquad = \underline{358.4 \text{ m/s}}$

Value is only approximate because the antinode is a little above the open end of tube. $\lambda/4$ is a little more than 35 cm.

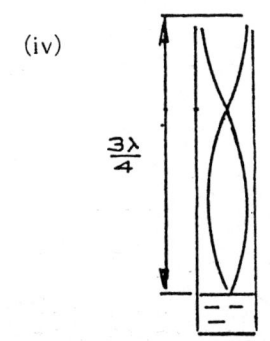

(iv) Second position of resonance occurs when air column length is $3/4\ \lambda$ (approx).

For A, $3/4\ \lambda = 105$ cm, therefore column length = 105 cm (not possible).

For B, air column lengths $\lambda/4$, $3\lambda/4$, i.e. 17.5 cm and 52.5 are both possible.

Question 17

PART I

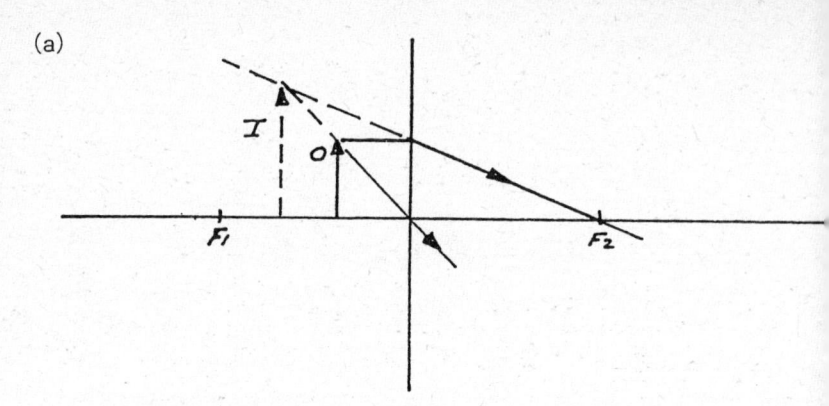

(a)

(b) As the object is moved along the axis to the left:-

   (i)   Nearer to - but not as far as - F1 the image moves further to the left and increases in size. It remains erect and virtual.

   (ii)   With the object at F1 the image is at infinity and, therefore, cannot be located. When the object is moved a finite distance beyond F1, the image is real, inverted and to the right of F2. Initially magnified, the image reduces in size as the object distance increases being the same size as the object when the object distance is twice that between the lens and F1.

PART II

(c) Focal length:- f = 10 cm, image distance:- v

   (i)   object distance:- u = 60 cm.

$$\frac{"1}{u} + \frac{1}{v} = \frac{1"}{f} \quad \therefore \quad \frac{1}{60} + \frac{1}{v} = \frac{1}{10} \quad \therefore \quad v = 12 \text{ cm}$$

$\therefore$ <u>The lens must be moved</u> (12-11) = <u>1 cm</u> to make image distance 12 cm.

(ii) with u = 300 cm and f = 10 cm we have:-

$\frac{1}{300} + \frac{1}{v} = \frac{1}{10}$ ∴ v = 10.34 cm

New image distance = 10.34 cm.

∴ <u>The lens must be moved towards the film a distance of</u> (12 - 10.34) = <u>1.66 cm.</u>

(d) (i) Aperture control:- this regulates the amount of light energy admitted via the lens. Together with the shutter, the aperture control provides the appropriate amount of film exposure.

(ii) When a fast moving object is being photographed, the shutter speed has to be adjusted (increased) in order to prevent a blurred photograph.

Question 18

(a)

PART I

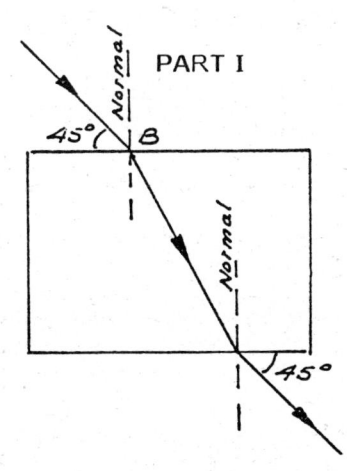

(b) As the ray enters the glass from the air it is bent towards the normal, i.e. is refracted. As the refracted ray leaves the glass and enters the air at the opposite face, it is refracted away from the normal through the same angle and continues along a path parallel to the original.

PART II

(c) velocity of light in air:- $\upsilon_A = 3 \times 10^8$ m/s
frequency of light:- $f = 6 \times 10^{14}$ Hz
refractive index, air to glass:- $_A\mu_G = 1.5$

(i) wavelength in air:-
$$\lambda_A = \frac{\upsilon_A}{f} = \frac{3 \times 10^8}{6 \times 10^{14}} = \underline{5 \times 10^{-7} \text{ m}}$$

(ii)

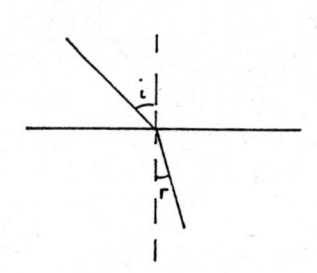

$$_A\mu_G = \frac{\sin i}{\sin r} \quad \therefore \sin r = \frac{\sin 45°}{1.5}$$

$$\therefore r = 28.13°$$

∴ The ray is turned through $(45 - 28.13) = \underline{16.87°}$

(iii) $_A\mu_G = \frac{\upsilon_A}{\upsilon_G}$ where υ_G = velocity in glass

$$\therefore \upsilon_G = \frac{3 \times 10^8}{1.5} = 2 \times 10^8 \text{ m/s}$$

wavelength in glass:- $\lambda_G = \frac{2 \times 10^8}{6 \times 10^{14}}$

$$= 3.3 \times 10^{-7} \text{ m}$$

(iv) Angle of incidence at lower face = angle of refraction:- r at upper face = $\underline{28.13°}$.

(d) If light of a lower frequency used:-

c(i) λ_A would increase as λ is inversely proportional to frequency.

c(ii) The angle through which the ray is turned would remain the same as it depends upon the refractive index of glass (in air). The ratio $\frac{\upsilon_A}{\upsilon_G}$ has not changed.

Question 19

PART I

(a) (i), (ii)

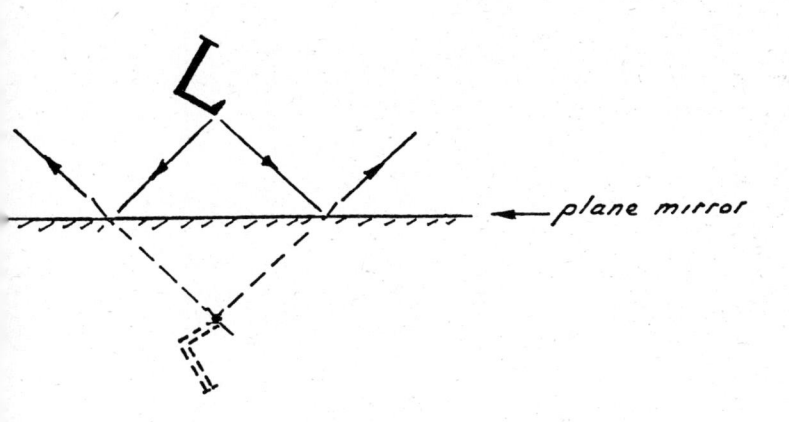

← plane mirror

(b) (i)   (ii)

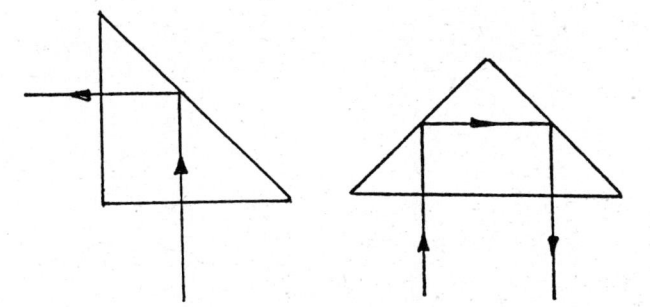

PART II

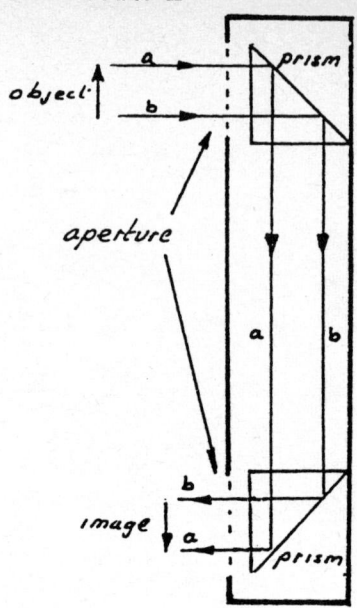

(c) (i), (ii)

object, a, b → *prism*

aperture

a, b

b, a → *image*, *prism*

(iii) The fault with this system is that it produces an inverted image.

(iv) The correction is made by using an erecting prism. The light is totally internally reflected at the base.

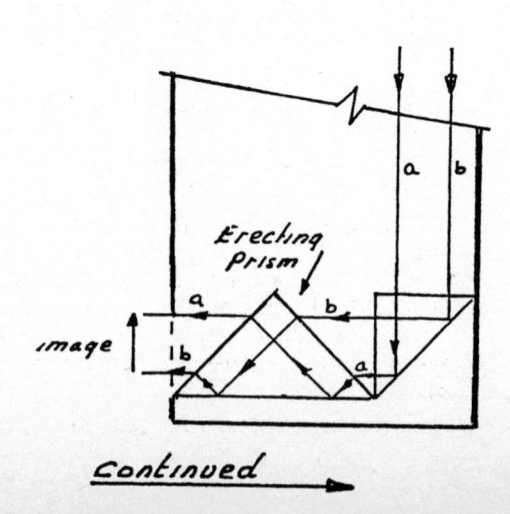

a b

Erecting Prism

image ← a, b

continued →

(d) <u>By calculation</u>:-

object distance = u , image distance = v

focal length = f = 200 mm

magnification = m = 1.5

$m = \dfrac{v}{u}$ ∴ v = mu = 1.5 u

"$\dfrac{1}{u} + \dfrac{1}{v} = \dfrac{1}{f}$" , ∴ $\dfrac{1}{u} + \dfrac{1}{1.5u} = \dfrac{1}{200}$

∴ $\dfrac{2.5}{1.5u} = \dfrac{1}{200}$

∴ u = $\dfrac{2.5 \times 200}{1.5}$ = 333.3 mm

<u>Object Distance = 333.3 mm</u>

<u>By Scale Drawing</u>:-

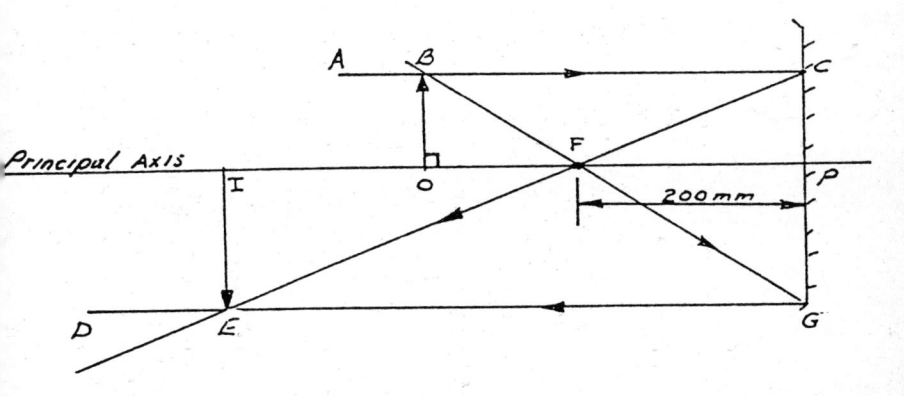

Method:- Draw the principal axis and, at distances in the ratio 2:3 from it, the parallel lines AC and DG. Draw GF - produced to meet AC at B. With BOP = 90⁰, OP represents the <u>object distance</u> which agrees, approximately, with the calculated value of <u>333.3 mm.</u> (IP represents the image distance = 1.5 × 333.3 = 500 mm)

Question 20

(a) (i) The incident ray is normal to the side AB and, therefore, passes into the prism without being refracted. The ray is totally internally reflected at side BC to meet side AC normally and to emerge from the prism without being refracted. As refraction does not take place at any of the sides, it is not possible to find a value for the refractive index of the glass by using this arrangement.

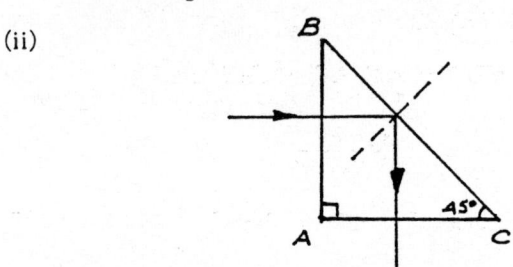

(ii)

(iii) Total internal reflection can occur only when light passes from a dense to a less-dense medium. The angle of refraction will increase to a maximum of $90°$ as the angle of incidence is increased and, at this point, all the light is reflected. For glass, the 'critical angle' of incidence is about $42°$ which has been exceeded in this case.

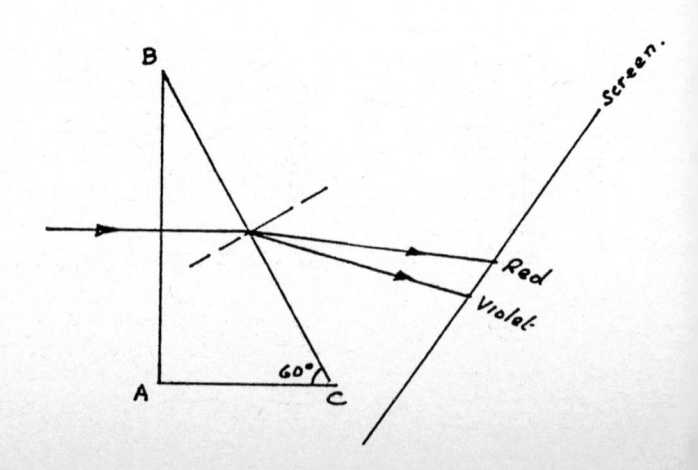

(b) (i)

The spectrum is formed on the screen because of differences in refractive index which occur for light of different wavelengths (colours). The deviation is least for red and greatest for violet.

(ii) If i = angle of incidence and r = angle of emergence

$$\mu = \frac{\sin r}{\sin i} \quad \therefore \quad 1.53 = \frac{\sin r}{\sin 30^0} \quad \therefore \quad r = 49.9^0$$

Angle of emergence = 49.9^0

(c) Refractive index = $\frac{\text{speed of light in air}}{\text{speed of light in glass}}$

$$\mu = \frac{V(\text{air})}{V(\text{glass})}$$

$$\therefore V(\text{glass}) = \frac{V(\text{air})}{\mu} = \frac{3.0 \times 10^8}{1.5} = 2 \times 10^8$$

$\therefore$ Speed of light in glass = 2×10^8 m/s

The change in speed:-

(i) has no effect on frequency

(ii) results in a decrease in wavelength.

Question 21

(a)

(b) magnification = $\frac{\text{distance of image from lens}}{\text{distance of object from lens}}$

$$= \frac{v}{u}$$

(b) magnification = $\dfrac{\text{distance of image from lens}}{\text{distance of object from lens}}$

$= \dfrac{v}{u}$

(c)

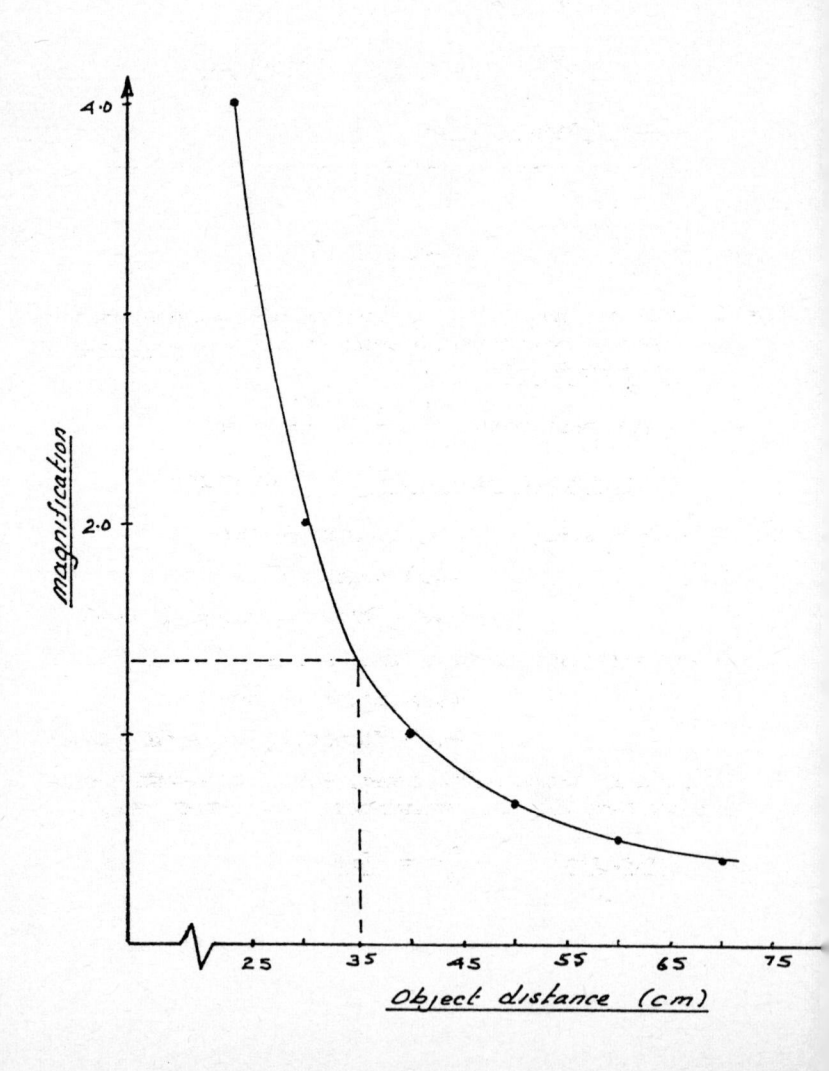

(i) The incorrectly determined magnification = <u>1.8</u>

(ii) The correct value of the magnification = <u>1.33</u>

(iii) When magnification = 1:-

object distance = image distance = 40 cm

from $\dfrac{1}{u} + \dfrac{1}{v} = \dfrac{1}{f}$,

$\dfrac{1}{40} + \dfrac{1}{40} = \dfrac{1}{f}$, ∴ f = 20 cm

<u>focal length = 20 cm</u>

Question 22

PART I

(a) (i) As A and B are in series, the current must be the same through each of them, ∴ <u>Current through B = 2A.</u>

(ii) Total resistance:- $R = \dfrac{V}{I} = \dfrac{12}{2} = 6\Omega$

∴ <u>Resistance of A = 3Ω</u> (A, B identical)

(b) A and B in series:- Total resistance = 6Ω

Total current = 8A

Power Output = VI = 24 watts

A and B in parallel:- Total resistance = 1.5Ω

Total current = 8A

Power Output = VI = 96 watts

Neglecting losses and heat capacities of the containers, temperature rise is proportional to power output.

∴ Temp. rise = $5 \times \dfrac{96}{24}$ = <u>20 K</u>

PART II

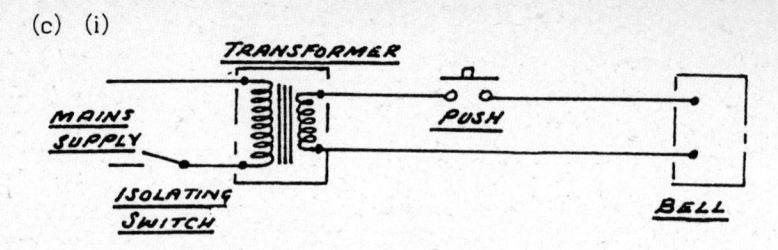

(c) (i)

[Diagram: Mains Supply connected through Isolating Switch to Transformer primary; secondary connected through Push to Bell]

(ii) Secondary e.m.f. = $240 \times \dfrac{90}{1800}$ = <u>12 volts</u>

(d) For X, voltage drop (secondary) = IR = 1.5 × 0.5
= 0.75 v
For Y, voltage drop (primary) = IR = 0.08 × 0.5
= 0.04 v

Thus:- Y's assertion is correct.

(e) X's method is safer because the bell push is located in the 12 volt circuit whilst in Y's method, the bell push is located in the 240 volt circuit which could be dangerous (e.g. in damp conditions) to the person operating the push.

Question 23

PART I

(a) (i) The resistor acts as a 'shunt' allowing some of the circuit current to by-pass the meter. By a suitable choice of value of resistance, full scale deflection can be obtained for currents greater than 10 mA provided, of course, 10 mA flows through the meter and the remainder through the 'shunt'.

(ii) <u>Resistance of the meter:-</u> $R = \dfrac{V}{I} = \dfrac{0.1}{0.01}$ = <u>10Ω</u>

(iii) For a full-scale deflection at 10 V we still require a meter current of 10 mA. A resistance - R - is connected in series:-

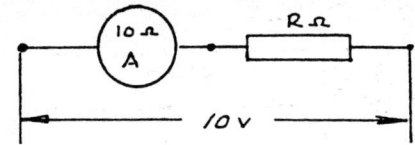

[Circuit diagram: 10 Ω ammeter A in series with R Ω, across 10 V]

Total Resistance = 10 + R

Current $I = \dfrac{V}{R} = \dfrac{10}{10 + R}$

but I = 0.01 A

∴ $\dfrac{10}{10 + R} = 0.01$

∴ 10 = 0.1 + R

∴ R = 999 Ω

PART II

(b) (i) Dry Leclanche cell has largest internal resistance
$= \dfrac{V}{I} = \dfrac{1.5}{1.5} = 1\,\Omega$

(ii) Internal resistance of wet Leclanche cell -
$R = \dfrac{1.5}{3.0} = \underline{0.5\,\Omega}$

(c) [Circuit diagram: E = 1.5 v, r = 0.5 Ω, R = 2 Ω, current I]

(i) Total Circuit Resistance = 2 + 0.5 = 2.5 Ω

(ii) Current = $I = \dfrac{E}{R + r} = \dfrac{1.5}{2.5} = 0.6$ A

(iii) **Terminal p.d.** = E - Ir = 1.5 - (0.6 × 0.5)
= 1.2 volts

(or, Terminal p.d. = IR = 0.6 × 2 = 1.2 v)

(d) A:- Step-down transformer:- 240V a.c. to 12V a.c.

B:- Diode:- allows current to flow only in one direction (in this case, B to C)

C:- Ammeter:- measures current flow.

D:- Rheostat (variable resistor):- allows current flow to be controlled.

Question 24

PART I

(a) (i)   (ii)

Viewed from above

(iii) In situation (ii), there is a force of attraction between the two wires.

PART II

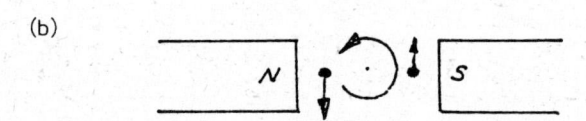

(b)

Forces acting on the two sides of the coil, indicated by vertical arrows. Direction of motion indicated by circular arrow. (Use Fleming's Left Hand Rule)

(c) When used as a generator:-

(i) Current flow is from F to E (INTERNALLY) i.e. from E to F (EXTERNALLY) (Use Fleming's Right Hand Rule)

(ii) The current produced is alternating. With the coil positioned as shown, maximum lines of force are being cut by the longer sides of the coil with the result that coil end at E is maximum positive and coil end at F is maximum negative. One quarter cycle later they are both at zero potential (no lines of force being cut), after another half cycle E is maximum negative and F maximum positive. This is indicated by the graph:-

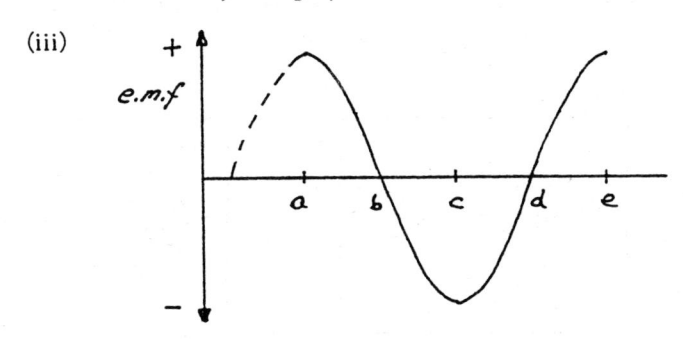

(iii)

[Graph: e.m.f vs time showing a sine wave with points a, b, c, d, e marked along the horizontal axis]

a → e represents 1 complete rotation of the coil in 1/4 revolution steps.

a:- starting point
b:- 1/4 revolution after start etc.

(d) (i) 50 Hz is the frequency of the supply, i.e. the number of complete cycles (as in (iii)) per second.

(ii) r.m.s. (root mean square) voltage is an average which takes into account the variation over a complete cycle. It is equal to the d.c. emf which produces the same heating effect when connected to a similar <u>resistive</u> circuit.

(iii) $V_{rms} = \dfrac{V_{max}}{\sqrt{2}}$ where V_{max} = max e.m.f.

Question 25

PART I

(a)

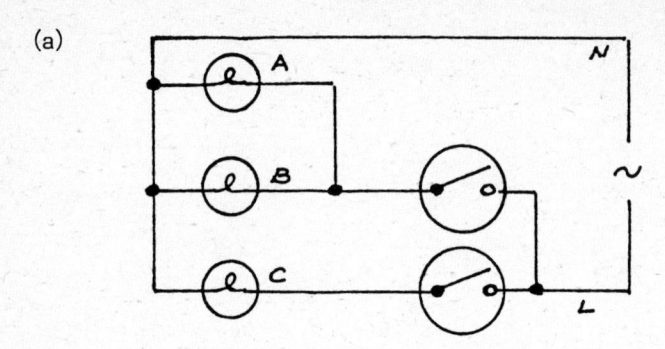

[Circuit diagram: Three lamps A, B, C in parallel between N and L lines, with two switches controlling lamps B and C, connected to AC source]

(b) Power = I^2R
    = $6^2 \times 0.01$
    = 0.36 W

<u>Energy</u> = Power × time
    = 0.36 × 5
    = <u>1.8 J</u>

PART II

(c) (i)

[Circuit: 60Ω and 120Ω in series with AC source]
$R = 180 Ω$
(LOW)

[Circuit: 60Ω and 120Ω in parallel with AC source]
$R = 40 Ω$
(HIGH)

[Circuit: 120Ω with AC source]
$R = 120 Ω$

[Circuit: 60Ω with AC source]
$R = 60 Ω$

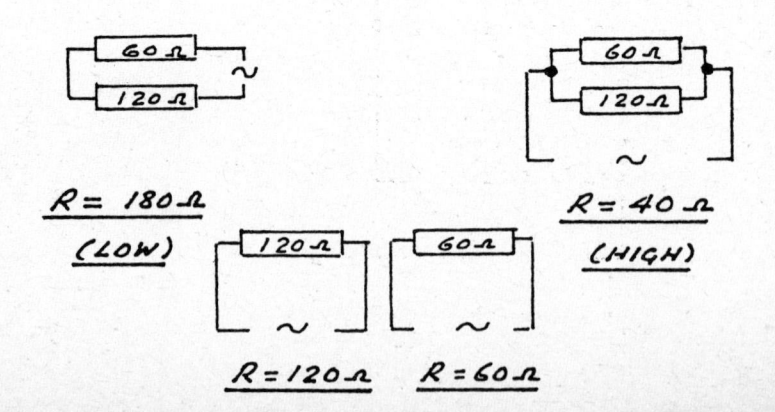

(ii) At max output (HIGH):-

Total Resistance $= \dfrac{1}{\dfrac{1}{60} + \dfrac{1}{120}} = 40\Omega$

with 240V a.c. supply,

Current $= I = \dfrac{V}{R} = \dfrac{240}{40} = \underline{6A}$

(iii) Let I = average current (amps) over 10 hr period

Power = VI = 240 I

Energy used in 10 hours $= \dfrac{240\ I \times 10}{1000} = 2.4I$ kWh

Cost = 2.4I × 5 = 12I pence

but cost = 48p

∴ 12I = 48

∴ I = average current = 4 amps

Question 26

(a) (i)

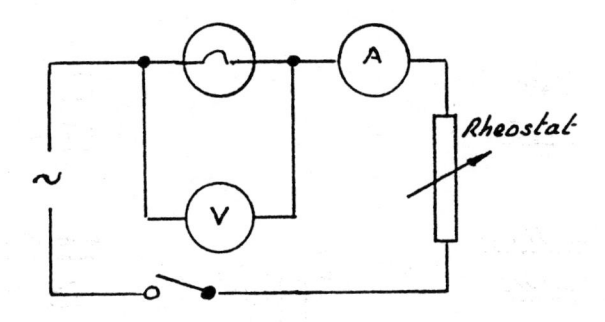

(ii) The current is varied by means of the rheostat, a variable resistor. As it is in series with the lamp, an increase in the resistance results in a current decrease - and vice-versa.

(iii) For each rheostat setting, readings are taken of voltage (v) and current (I) from V and A.

Resistance:- $R = \dfrac{V}{I}$ ohms.

(iv) For a metal tungsten filament, the resistance would be expected to increase as the current is increased. This is because filament temperature is greater with increased current and, for most metals, resistance increases as temperature increases.

(b) (i) Initially, the filament temperature - and resistance - is relatively low and, without a thermistor, the current would be high enough to damage the delicate filament. The inclusion of the thermistor limits the initial current to a safe value. As the thermistor and filament temperatures increase with current flow, their resistances decrease and increase, respectively. There will be an increase in current until it reaches its normal working maximum.

(ii) When working temperature and current have been attained, the thermistor has no further function and is 'short-circuited' through S2. However, S2 must be opened again when S1 is opened to ensure that the thermistor is in circuit when the lamp is again switched on from 'cold'.

Question 27

(a) (i) If the magnet is lowered quickly into the coil there is a deflection on the galvonometer for the very short time that the magnet is moving. Current flow is such that the resulting magnetic field opposes motion, i.e. a N-pole at top of coil:- G is deflected to the left.

(ii) No relative movement between coil and magnet, therefore no deflection.

(iii) With magnet moving in - deflection to the left.
With magnet moving out - deflection to the right.
Thus:- the induced e.m.f. is alternating with a frequency of, about, 2 cycles per sec. = 2 Hz.

(b) When connected to the a.c. mains, there is an alternating current flowing through the primary coil which produces an alternating magnetic field throughout the soft-iron core. As a consequence, an e.m.f. is induced in the secondary coil. As the magnitude of the field produced in the primary and the e.m.f. induced in the secondary is proportional to their number of turns, we have:-

$$\frac{\text{Secondary e.m.f.}}{\text{Primary e.m.f.}} = \frac{\text{No. of Secondary turns}}{\text{No. of Primary turns}}$$

∴ in this case:- Secondary voltage = $240 \times \frac{5}{400}$

$$= 3 \text{ volts}$$

(i) The nail melts because although the voltage across it is low (3v), the resistance is very low and resulting current is very high.

(ii) Laminations reduce the formation of energy losses due to eddy currents which are set up in the core.

(iii) If the core is incomplete, the magnetic circuit is broken and there is no flux linking with the secondary coil. Hence:- no voltage across the nail to provide the current.

Question 28

Absorber material	Ratemeter reading counts per minute			
	1	2	3	mean
Air	120	110	130	120
Paper	100	120	110	110
Cardboard	130	130	100	120
Aluminium (0.5 mm)	110	120	110	113⅓
Aluminium (5 mm)	50	60	40	50
Lead (5 mm)	40	50	50	46⅔
Lead (50 mm)	50	40	50	46⅔

(a) The mean counts per minute for each absorber material is indicated in the above table.

There is no significant difference in the mean counts per minute for the absorber materials:- air, paper, cardboard and 0.5 mm thick aluminium. α-radiation will not penetrate paper so it follows that α-radiation is not present. β-radiation will penetrate 0.5 mm but not 5 mm aluminium, therefore the significant decrease in counts with the latter material would indicate the presence of β-radiation. If γ-radiation were present, it would be stopped by 50 mm lead. As there is no decrease in counts at this stage it may be concluded that only β-radiation is present.

(b) The absorber could be replaced by an electro-magnet arranged so that the radiation passes between the poles. Arrangement of the G-M tube for maximum count would indicate the emission of negatively charged particles:- β-radiation.

(c) Ratemeter readings for the 50 mm lead absorber are due to emission from radioactive material in the earth and in nearby surroundings. This "background count" would be evident in the absence of any provided experimental source.

(d) The difference in readings arises because of the random nature of radioactive emission.

(e)

Absorber material	Ratemeter reading counts per minute	
	α-source	γ-source
Air	120	120
Paper	50	120
Cardboard	50	120
Aluminium (0.5 mm)	50	120
Aluminium (5 mm)	50	120
Lead (5 mm)	50	120
Lead (50 mm)	50	50

Question 29

(a) Electron gun:- Electrode K is the cathode which provides a supply of electrons by thermionic emission. Electrode L is a cylindrical anode, maintained at a high positive potential, which accelerates the electrons before they emerge as a beam.

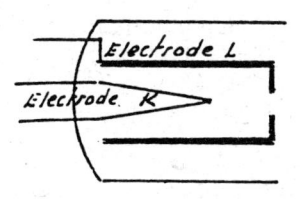

The purpose of the coils is to provide a means by which the electron beam may be deflected in the horizontal plane. The deflection is achieved by passing a current through the coils, the direction and magnitude of which will determine the direction and magnitude of the resulting magnetic field and deflection.

(b) See following page.

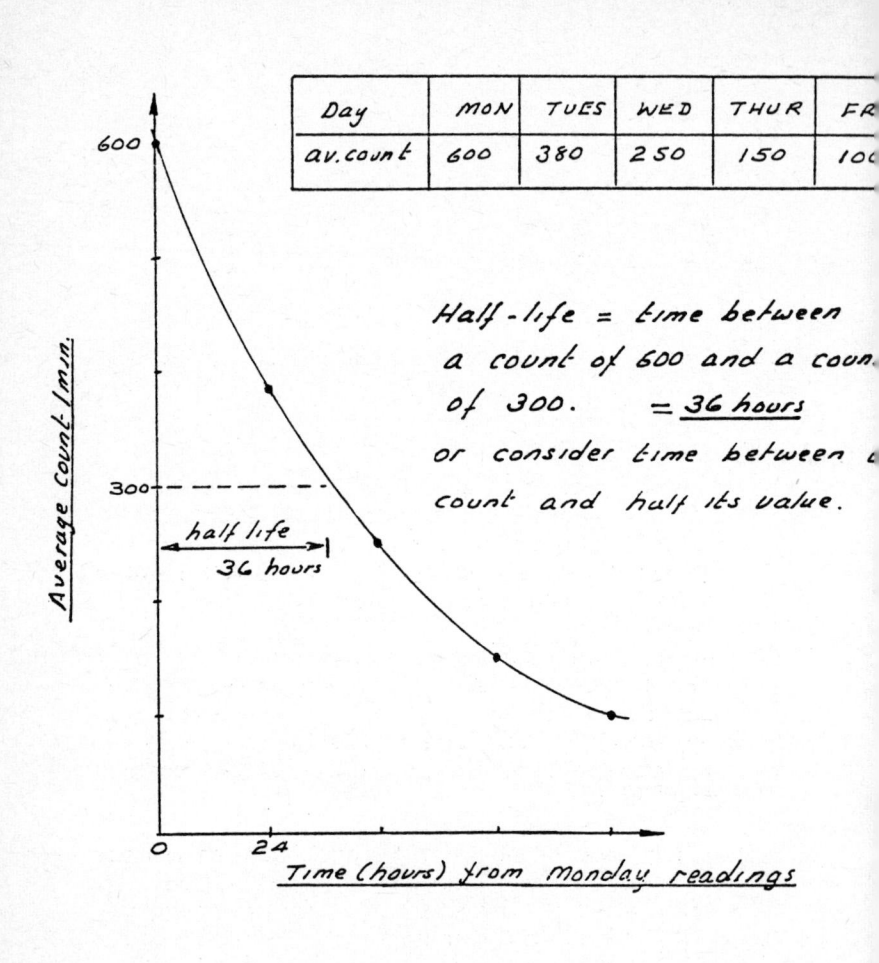

(b)

Day	MON	TUES	WED	THUR	FR
av. count	600	380	250	150	100

Half-life = time between a count of 600 and a coun. of 300.   = 36 hours

or consider time between a count and half its value.

Average count/min. — vertical axis, marked 600 and 300
Time (hours) from monday readings — horizontal axis, marked 0, 24
half life 36 hours

Question 30

(a)

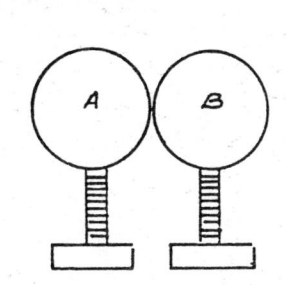

uncharged - in contact

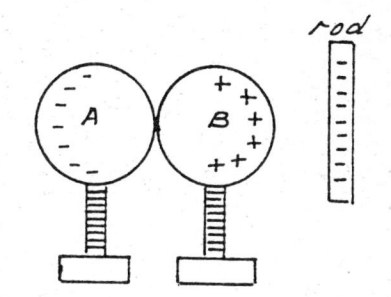

+ve charge induced on B
−ve charge induced on A.

(i) A and B are placed - uncharged - in contact and the negatively charged rod is brought near to B. A positive charge is induced on B and an equal negative charge on A. A is separated from B. A and B remain charged.

(ii) A gold leaf electroscope can be used to test the charges on A and B. When positively charged and B is brought near there is an increase in divergence. Similarly, A produces the same effect when brought near to a negatively charged electroscope.

(iii) Being good conductors, A and B contain a large number of free electrons. When in contact, they may be regarded as a single conductor. With the rod in proximity to end B, free electrons are repelled towards A thus giving A a net negative charge and leaving B with an equal net positive charge. On separation, these charges remain.

(b) (i) "$I = V/R$" :- Current = $\dfrac{2}{5 \times 10^6}$ = 4×10^{-7} Amps

(ii) "$Q = It$":- Charge = $4 \times 10^{-7} \times 1 = 4 \times 10^{-7}$ C

(iii) Charge carried by one electron = 1.6×10^{-19} C

$$\therefore \underline{\text{No. of electrons}} = \frac{4 \times 10^{-7}}{1.6 \times 10^{-19}} = \underline{2.5 \times 10^{12}}$$

THE BASIC CONCEPTS SERIES

The Basic Concepts series attempts to explain in a clear and concise manner the main concepts involved in a subject. Paragraphs are numbered for ease of reference and key points are emboldened for clear identification, with self assessment questions at the end of each chapter. The texts should prove useful to students studying for A level, professional and first year degree courses. Other titles in the series include:—

    Basic Concepts in Business by Tony Hines
    Basic Concepts in Foundation Accounting by Tony Hines
    Basic Concepts in Financial Mathematics and Statistics
        by T.M. Jackson
    Basic Concepts in Business Taxation by K. Astbury

QUESTIONS AND ANSWERS SERIES

These highly successful revision aids contain questions and answers based on actual examination questions and provide fully worked answers for each question. The books are written by experienced lecturers and examiners and will be useful for students preparing for O and A level, foundation and BTEC examinations. Subjects include:—

    Economics by G. Walker
    Accounting by T. Hines
    Multiple Choice Economics by Dr. S. Kermally
    O level Mathematics by R.H. Evans
    A level Pure Mathematics and Statistics by R.H. Evans
    A level Pure and Applied Mathematics by R.H. Evans.
    O level Physics by R.H. Evans
    O level Chemistry by J. Sheen
    O level Human Biology by D. Reese